THE GREENHOUSE TRAP

The Greenhouse Trap

WHAT WE'RE DOING TO THE ATMOSPHERE AND HOW WE CAN SLOW GLOBAL WARMING

FRANCESCA LYMAN

WITH IRVING MINTZER, KATHLEEN COURRIER, AND JAMES MACKENZIE

BEACON PRESS   BOSTON

World Resources Institute is deeply grateful to the Florence and John Schumann Foundation and the Curtis and Edith Munson Foundation, Inc., for support for WRI Guides to the Environment.

Beacon Press
25 Beacon Street
Boston, Massachusetts 02108-2892
www.beacon.org

Beacon Press books
are published under the auspices of
the Unitarian Universalist Association of Congregations.

First digital-print edition 2001

ISBN 0-8070-8503-0
LCN 89-43080

CONTENTS

ABOUT WRI GUIDES TO THE ENVIRONMENT

Every so often there is a sea change in world affairs, and we're amid one now. Environmental concerns are back on the front page and in the nightly news. Even more important, they are being discussed at both the dinner table and the international negotiating table. It has slowly dawned on Americans that waste, pollution, congestion, land degradation, and the like are not isolated mishaps that time will cure, but are instead by-products of the way we live. And with this recognition has come another: that we have choices to make—the sooner, the better.

We can't make those choices, of course, without a firm fix on the facts and some sense of what's economically and politically possible in our own lifetimes. And we also need a shared understanding of how our busy, productive, big-spending, and still widely emulated country shapes the planet's environmental future and—the flip side—of how economic and social choices in the rest of the world influence us.

To help Americans grasp the big picture, World Resources Institute created *WRI Guides to the Environment*. These books were conceived with a sense of urgency to dispel confusion about the greenhouse effect, the loss of the rain forests and biological diversity, environment and development in Third World countries, energy alternatives, and other concerns that newspapers and television usually cover piecemeal or in quick provocative "takes" that leave us either baffled or discouraged. Our belief is that even the most scientifically complex environmental issues can be clearly explained, that there is plenty that we can all do about them, and that Americans are ready to try.

Kathleen Courrier and Mohamed T. El-Ashry
Series Editors

FOREWORD

Climate change isn't just any environmental issue. *It's bigger.* Scarcely a life, much less a country, can escape its effects. *It's virtually irreversible.* The healing time must be reckoned in centuries or millennia, not lifetimes. *It's tied to almost every facet of contemporary economic life.* How we travel, manufacture and ship goods, build buildings, farm, and spend our leisure time all influence the tempo of climate change. *And if it's not already here, an altered climate is probably coming soon.*

Probably? Yes, that's right: probably. One unusual feature of this environmental threat is that scientists can't say with certainty that climate change is actually occurring yet or what its precise effects will be. What they can say is that the average global temperature has already risen about 1 degree Fahrenheit in the last century, compared to a natural rate of less than one degree per millennium over the last 12,000 years. They know that the "greenhouse effect" is real: if certain gases already naturally present in the atmosphere, such as carbon dioxide, didn't have the ability to trap heat, Earth would be a sterile, iced-over orb. And they know that these gases and other so-called "greenhouse" gases have been rapidly accumulating in the atmosphere in recent decades due to human activities. These "banked" emissions could warm Earth by a total of 0.9 to 3.2 degrees Fahrenheit no matter what we do to control emissions from here on out.

To some scientists and others, the available evidence is cause enough to clang the warning bell and announce that the global warming has begun. Others are reserving judgment until more measurements can be taken, more data gathered, more sophisticated computer models built. In the last year or two, some "undecideds" have drifted into the camp of the believers, but as of late 1989 scientists remained divided.

Surprisingly, this standoff is not the end of the story, but the beginning. Even experts who aren't convinced that rapid climate change is upon us are quite convinced that it will cause momentous problems if it comes. And it's but a half step from this understanding to the realization that we should immediately do what we can to make sure that the next one-degree temperature rise is not followed quickly by another and another.

In many countries, the growing sense that something must be done defies the political system's built-in tendency to stall until crisis presents itself ugly and full-fledged. Who knows? Maybe the recent thaw in U.S.– Soviet relations has allowed at least the superpowers to look anew at what really poses the most imminent threat to human survival.

How far the United States and others will go—how much, in particular, we will spend to retool our energy system—without 100 percent certainty is one of the most important questions we face as the new century draws near. If we don't act quickly and creatively now, the question for the next generation may be even more difficult: how can we preserve individual freedom if our life-support systems are threatening to collapse all around us?

Fortunately, the measures needed to slow climate change dovetail with those needed to solve a wide range of problems already on public agendas—urban smog, acid rain, insecure energy supplies, tropical deforestation, ozone-layer depletion, and others. With so many reasons to act *besides* the compelling ones related to climate change itself, there is hope that something can be done in time to keep Earth on an even keel.

Even though climate change threatens to be bigger, more irreversible, and more pervasive than other environmental problems, it's also *controllable* if we act now. That much is clear.

Gus Speth, President
World Resources Institute

ACKNOWLEDGMENTS

We would like to acknowledge the time and care of many people in helping research the material for this book. We have drawn particularly on the resources of the Energy Conservation Coalition, Alliance to Save Energy, American Council for an Energy Efficient Economy, Environmental Action, Friends of the Earth, Conservation Law Foundation, Lawrence Berkeley Laboratories, Renew America, Earth Day 1990, the Climate Institute, the American Forestry Association, California ReLeaf, and Worldwatch Institute. Indirectly, *Harpers* magazine also contributed by first coming up with the format we use in the first section of *The Greenhouse Trap*.

For reading and criticizing drafts of the manuscript, we thank Tolak Besman, Antonia Sherman, Regan Heiserman, Rosina Bierbaum, Kevin Finneran, Laura Tangley, Deborah Johnson, and the League of Women Voters of Bronxville, N.Y.

We would also like to thank the staffs of the Yonkers and Bronxville Public Libraries (especially Patricia Root), the Sarah Lawrence College Library, the Concordia College Library, and the WRI Library for their help and advice. Thanks to Maia Garrison, Affifa Wasely, Estelle Aglow, Amber Leonard, Sue Terry, and Bob Aglow for their unending support.

Our biggest debt, of course, is to World Resources Institute. WRI's president, Gus Speth, put climate change on the institute's agenda nearly seven years ago, and we believe that he has also been instrumental in getting other organizations and governments to do the same. We would also particularly like to thank Don Strandberg for planting a seed that led to this series, Bill Moomaw for his careful reading, John Ceccatti for his able help preparing tables and other special material, Jeannie Oliver and Moira Connelly for helpful research assistance, Hyacinth Billings for production advice, Sheila Mulvihill for copyediting, and Maggie Powell, Lori Pierelli, and Claudia

Bedwell for patiently retyping drafts. Reviews by Walt Reid and Mark Trexler of WRI also proved invaluable.

Much of World Resources Institute's success in raising public consciousness of the risk of rapid climate changes is due to the efforts of Rafe Pomerance, who had the "greenhouse fire," as he called it, before he moved to WRI from Friends of the Earth. We owe Rafe a special thank you for educating us to the issue and keeping us abreast of many aspects of the issue over the years. Thanks also go to John Hoffman, Dennis Tirpak, Steve Seidel, and Jim Titus of EPA; Alan Miller of the University of Maryland; and John Topping of the Climate Institute.

We owe special thanks to Mohamed T. El-Ashry, who provided the guidance needed in our work and the vision to sustain the entire series. This book would not have been possible without him.

F. L., I. M., K. C., and J. M.

CLIMATE CHANGE FACTS

How We Use Our Energy

• *Average number of gallons of fuel consumed by each U.S. car in 1967: 715*

• *Average number of gallons of fuel consumed by each U.S. car in 1987: 515*

• *Number of new trucks (all sizes) and cars bought in the United States in 1967: 9,900,000*

• *Number of new trucks (all sizes) and cars bought in the United States in 1987: 15,000,000*

• *Number of new light trucks (under 14,000 pounds) bought in the United States in 1987: 4,600,000*

• *Number of megawatts of domestic hydroelectric power the United States had the potential to produce in 1984 if dams were built in every physically possible location: 183,000*

• *Number of megawatts of hydroelectric power actually produced in the United States in 1985: 88,000*

• *Percentage of oil used in the United States in 1988 that was imported: 38*

• *Percentage increase in use of electricity by U.S. homes and businesses, 1973–88: 64*

• *Percentage of energy used in the United States that comes from fossil fuels: 90*

• *Percentage of oil used in Western industrialized countries that is consumed by U.S. cars: 13*

• *How much less efficient light trucks are than cars: 1/3*

- *Percentage of light trucks used mainly to drive to work or to the shopping mall: 75*

- *Percentage of scientists and engineers who think that nuclear energy plants will be "very or somewhat important" to meet U.S. electricity needs: 83*

- *Billions of dollars the United States spent in 1988 on imported crude and petroleum products: 39*

- *Percentage of the world's electricity supplied by hydropower in 1986: 21*

- *Percentage of power cut in a brownout: 5*

How Energy Use Affects Us and the Environment

- *Number of gallons of oil dumped into Prince William Sound by an Exxon tanker in 1989: 11,000,000*

- *Number of pounds of carbon dioxide emitted into the air for each mile driven in a passenger car that gets 22 miles per gallon: 0.9*

- *Number of pounds of carbon dioxide produced by burning 1 gallon of gasoline: 19*

- *Number of pounds of carbon dioxide produced by using one kilowatt-hour of electricity from a coal-fired plant: 2*

- *Number of miles it takes for each airplane passenger to generate 1 pound of carbon dioxide: 2*

- *Number of tons of carbon dioxide each American generates in a year: 20*

- *Number of passenger cars in the United States in 1988: 140,000,000*

- *Number of tons of carbon dioxide that a car getting 27.5 miles per gallon emits over its ten-year lifetime: 35*

- *Number of Americans who live in areas that exceed Environmental Protection Agency ozone pollution standards: 80,000,000*

- *Rank of Hawaii, Vermont, and Wisconsin for states with the* highest *estimated miles per gallon for cars: 1, 2, 3*

- *Rank of Arkansas, Arizona, and Louisiana for states with the* lowest *estimated miles per gallon for cars: 1, 2, 3*

- *Percentage of skin disease among fair-skinned people caused by the sun: 50*

- *Number of people who die from skin cancer in the world each year: 100,000*

- *Rank of the United States, USSR, and Western Europe as leading industrial emitters of carbon dioxide: 1, 2, 3*

WHAT WE'RE DOING ABOUT IT

- *Average improvement in miles per gallon for new cars required by 1991 to meet auto fuel-efficiency regulations: 1*

- *Percentage of people polled by* Parents *magazine on the greenhouse effect who said they would be willing to drive less to help solve the problem: 69*

- *Who would take fewer showers: 67*

- *Who would take fewer airplane trips: 61*

- *Percentage of Americans willing to see fuel economy standards raised to 45 miles per gallon by the year 2000—even if it adds $500 to the cost of a car: 77*

- *Number of trees Applied Energy Services will plant to offset the carbon their 183-megawatt coal-fired power station will emit: 52,000,000*

- *How many times longer an 18-watt compact fluorescent bulb will light a room compared to a 75-watt incandescent bulb: 13*

- *Millions of dollars E.I. DuPont de Nemours & Company spent in 1988 to find a substitute for chlorofluorocarbons: 70*

- *Percentage your water heating bills will be cut for each 10-degree reduction in the water heater setting: 3–5*

- *Number of states that fund ride-sharing programs: 16*

- *Number of states that have high-occupancy–vehicle lanes: 9*

- *Percentage of energy saved when using recycled aluminum as opposed to using primary metal: 95*

BECOMING A GLOBAL HOTHOUSE

- *Number of times on record July has been hotter in Washington, D.C., than it was in 1987 and 1988: 0*

- *The number of degrees Fahrenheit the average temperature has changed since the Little Ice Age: 2*

- *How much lower the coldest global average temperature was during the last Great Ice Age 15,000 years ago: 9 degrees Fahrenheit (5 degrees Celsius)*

- *The world's average temperature in degrees Fahrenheit: 59*

- *The warmest two years on global record: 1987 and 1988*

- *Of the six warmest years of the past century, the number that have occurred in the 1980s: 6*

- *Inches the sea level is expected to rise by 2025 due to current emissions of greenhouse gases: 5–15*

- *Average number of days per year in which the temperature of New York City exceeds 90 degrees Fahrenheit (32 degrees Celsius): 15*

1

Is Earth Really Getting Warmer?

A growing scientific consensus exists that
sometime in the next century, the surface of
the earth will become warmer than it has been
at any time in human history.

> JOHN FIROR
> Deputy Director
> National Center for Atmospheric Research

Whilst a warming of this amount is consistent
with the "greenhouse effect" caused by the rise
in the levels of carbon dioxide and other
man-made gases in the atmosphere, the earth's
temperature fluctuates considerably due to
natural causes and no unambiguous connection
can yet be made.

> DAVID PARKER
> British Meteorological Office

Imagine a long, hot summer—longer, hotter, and stranger than any earthly season experienced today. In some regions, spring, fall, winter, and summer all merge into the hottest season within memory. Imagine days of searing 100-degree heat, beastly nights, bad air, parched fields, wildcat forest fires.

Human history contains no analog for the climatic conditions into which a "greenhouse effect" could plunge Earth. We must go back to the time when dinosaurs roamed the planet, several million years ago (when temperatures ranged from as much as 9 to 16 degrees higher than today's), or forward to science fiction, to glimpse such "hothouse" worlds.

We can also compare these visions to other planets in our solar system. Twenty-six million miles from Earth, in an orbit much closer to the sun, Venus spins through space with a furnacelike surface temperature of over 800 degrees Fahrenheit—much hotter than its proximity to the sun alone would make it. Scientists agree that the planet fell victim to a runaway "greenhouse effect," as the heat-trapping gas carbon dioxide somehow took over its atmosphere and rendered it lifeless.

Unlike its hellish sister planet, Earth is blessed with an atmosphere rich in nitrogen and oxygen, with only relatively small traces of carbon dioxide (only 0.03 percent, compared to 98 percent for Venus). But Earth's atmosphere has been transformed during the last century, as our society has pumped into it billions of tons of carbon dioxide and large amounts of other gases that absorb the heat energy emitted from the planet's surface. These gases are trapping enough thermal radiation to warm the globe by several degrees; and, if current trends continue, the buildup of these gases will warm the planet further, perhaps by as much as 8 or 9 degrees Fahrenheit by the middle of the twenty-first century. Even if we stopped adding to the buildup now, the increase that has already occurred could warm Earth by 0.9 to 3.2 degrees.

No one suggests that Earth will suffer a fate akin to that of Venus. But scientists do fear that a global warming greater than any known to humankind might occur and that it would do so at an unprecedented speed (within a century), taking a catastrophic toll on human societies and natural ecosystems.

Not since the beginning of written history some 8,000 years ago was Earth even 4 degrees Fahrenheit warmer than it is today. The only change of such magnitude on record in human history was in the other direction—cooling. Historian Barbara Tuchman described the impact of the Little Ice Age six centuries ago (when it was at most 3 degrees colder than today) in *A Distant Mirror: The Calamitous Fourteenth Century*:

> A physical chill settled on the fourteenth century at its very start, initiating the miseries to come. The Baltic Sea froze over twice, in 1303 and 1306–7; years followed of unseasonable cold, storms, and rains, and rise in the level of the Caspian Sea. Contemporaries could not know it was the onset of what has since been recognized as the Little Ice Age. . . .

Scenarios written by scientists—not science fiction writers—now suggest that even more devastating changes could be in store for the twenty-first century. The warming could seriously disrupt the weather, altering rain and snow patterns, shifting growing seasons, and possibly disturbing air and ocean currents enough to produce more intense and frequent hurricanes and tornadoes than plague us now. The melting of land glaciers and the gradual thermal expansion of seawater could raise sea levels enough to inundate coasts and make today's shoreline maps relics by the third quarter of the next century.

If the world heats up as much as some scientists now fear, the change will alter virtually every facet of contemporary life, playing havoc with farming, forestry, fishing, transportation, water supplies, and energy. Even military strategy might be affected; certainly, Pentagon officials and other national security experts will have to develop contingency plans to deal with the prospect of millions of new environmental refugees.

In the fierce heat of the world's hottest years on record—1987 and 1988—and the drought of 1988, warnings of a global warming took on a particularly striking significance. After years of routine science reporting on a possible greenhouse problem in the future, newspaper and TV reports were saturated with coverage of scientific predictions. Headlines read: "Feeling the Heat," "The Heat *Is* On," and even "Is the World Coming to an End?"

These headlines were matched by other pressing environmental issues. There were reports of beaches strewn with medical waste and of obsession with the weather. Suddenly, people were reading the worst-case scenarios of the twenty-first century. The late 1980s saw report after report on the ills of climate change, with visions of dying trees, drought, pestilence, and famine.

In June 1988, climate change became "official" to most Americans. Atmospheric scientist James Hansen of NASA's Goddard Institute for Space Studies told a U.S. Senate committee that the most searing heat and drought of this century were just a warm-up for what's to come. Shocking the public, he claimed to be "99-percent confident" that the continued rise in global temperatures over the last decade was not a random event, but a real sign of the predicted warming trend. "The greenhouse effect," he proclaimed, "has been detected and it is changing our climate now."

Besides capturing public attention, these events galvanized political will, leading many to conclude that the scientists who warned that we may be changing our climate unalterably were right. As Norway's

Prime Minister, Gro Harlem Brundtland, told The Toronto Conference on the Changing Atmosphere in June 1988: "While theories about the physical effect of carbon dioxide on the climate were presented more than a hundred years ago, what is new is the certainty that it will happen unless we take decisive corrective action now!" By summer's end, a campaigning George Bush called 1988 "the year the Earth spoke back" and promised to combat the greenhouse effect with a "White House effect."

In the cooler months that followed, there were conflicting reports. Some scientists presented evidence suggesting that the drought was not necessarily connected to the much-vaunted global warming. No, they said, the summer's heat was instead a manifestation of natural climatic cycles; specifically, shifts in the oceans and in the jet streams in the upper atmosphere were the cause. Researchers at the National Oceanic and Atmospheric Administration reported that they had found no evidence of a global warming trend in the century-long records of U.S. average annual temperatures. Many then wondered how seriously the warnings of a "greenhouse effect" were to be taken. Was the issue overplayed because of flukishly warm weather? Had the media sounded a false alarm?

The answer, according to many scientists, is no. Whether or not the air feels warmer or cooler from one summer to the next, whether or not one year's drought is part of a warming already under way, most scientists and experts believe that, if current trends in greenhouse gas emissions continue, the planet will undergo a long-term warming. The consequences of an anticipated warming are vividly stated in a 600-page U.S. government report released in 1988. According to the U.S. Environmental Protection Agency (EPA), the warming would degrade most of the spectrum of natural ecosystems and affect "when, where, and how we farm; the availability of water to drink and to run our factories; how we live in our cities; the wetlands that spawn our fish; the beaches we use for recreation; and all levels of government and industry."

Conducted at the request of Congress, the EPA study draws upon 54 research projects by government and academic scientists on coastal resources, water, forests, agriculture, biological diversity, air pollution, electric power demand, and human health. According to EPA, the most immediate impacts would probably be felt in agriculture and forestry, particularly among species sensitive to temperature rise. Greatly decreased forest yields, crop losses, and "moving" grain belts are among the predictions. The EPA findings "collectively suggest a world that is different from the world that exists today." The landscape would change in ways that couldn't be "fully predicted," the agency reports, but the resulting impacts would be "irreversible." Other recent research from around the world confirms this judgment.

INTEMPERATE ZONES

Although the *theory* of the "greenhouse effect"—that gases in the atmosphere trap heat close to Earth and redirect some of that heat back to Earth's surface—has been accepted for many years, a number of *facts* now make the global warming much more certain. The total planetary temperature has jumped 1.0 degree Fahrenheit during the last century, leaving the temperature in the 1980s higher than at any time since it was first systematically measured 130 years ago. Levels of carbon dioxide have risen 25 percent since preindustrial times, and they are increasing by an estimated 1.4 to 2.5 parts per million each year (0.4–0.7 percent annually), according to the National Academy of Sciences. Other greenhouse gases are also building up in the atmosphere.

This warming trend shows no sign of letting up. Data gathered around the world indicate a more rapid warming rate in the 1980s than in any earlier period. In 6 out of the last 10 years, average global temperatures were the highest on record during the last century, according to figures released in early 1989 by scientists with the Meteorological Office of the United Kingdom and the University of East Anglia. These temperature increases have been occurring despite natural factors—low levels of solar radiation, high levels of volcanic activity, shading, and increased dust and sulfate particulates in the air—that keep the world cooler than it might otherwise have been.

Even as the facts accumulated, a string of climate anomalies in 1987 and 1988 in the United States and around the world also left some indelible *impressions* of how a global warming might feel since part of the effect such a climate change would have on Earth is to make weather patterns more unpredictable:

- The Midwest experienced a devastating drought in 1988, as hot, dry weather parched crops, particularly corn and soybeans. During 1989, soil moisture levels were unusually low despite the return of winter rains. As a result, yields of winter wheat were down by 40 percent.

- Record heat that year in the western states of Wyoming, Idaho, Montana, and California created conditions that permitted a rash of forest fires, and vast areas of Yellowstone Park burned up in blazes triggered by lightning and carelessness.

- In fall 1988, Hurricane Gilbert, the most forceful hurricane ever recorded, swept across the Caribbean Sea and devastated communities in Jamaica, Mexico, and southern Texas.

- In 1987, a heat wave killed 1,200 people in Greece. Extreme heat and a late monsoon in India in 1988 killed hundreds there. Droughts, floods, and landslides in south and east China claimed crops and some 11,500 lives. And in the Sudan, after regional droughts lasting two years, a whole year's worth of rain fell in three days during August 1988, leaving 1.5 million people homeless.

- One of the worst floods in the last 100 years inundated Bangladesh in 1988, washing out many homes and causing thousands of deaths and billions of dollars worth of damage. Decades of deforestation and increasing soil erosion in the foothills, upstream from Bangladesh, reduced local watersheds' ability to absorb the rains, thus intensifying the flood damages.

WHAT CAUSES THE GREENHOUSE EFFECT?

The greenhouse effect has been an essential part of Earth's history for billions of years. Natural background concentrations of carbon dioxide and water vapor in the air have warmed the planet from about 0 to 59 degrees Fahrenheit. This warming allowed water to exist as a liquid in the oceans, streams, and lakes that cover six-sevenths of Earth's surface. Without liquid water, life as we know it would not have evolved.

Since the Industrial Revolution, this age-old balance has changed. Today, the presence of additional quantities of greenhouse gases threatens to make our simple "greenhouse" look far more ominous than its name implies. In particular, five important trace gases are combining to amplify the natural greenhouse effect:

- *Carbon dioxide* emissions from human activities currently account for about half the anticipated temperature increase. It is produced mostly by burning such fossil fuels as natural gas, coal, and petroleum, and

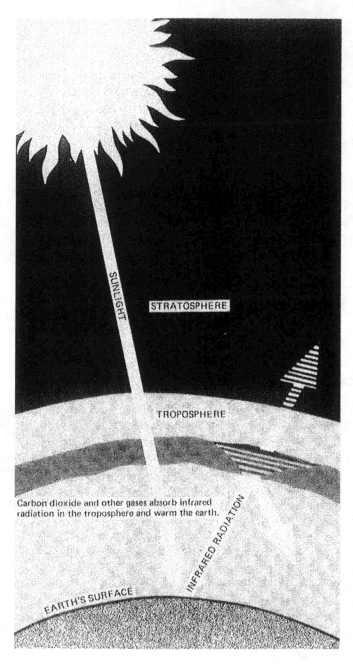

SUNLIGHT

STRATOSPHERE

TROPOSPHERE

Carbon dioxide and other gases absorb infrared
radiation in the troposphere and warm the earth.

INFRARED RADIATION

EARTH'S SURFACE

How the greenhouse effect works

by burning wood. About 80 percent of all global emissions come from fossil fuel combustion; the rest stem from deforestation (as trees and vegetation are burned or cut and allowed to decay). The concentration of carbon dioxide has increased by 10 percent just since 1958, by 25 percent since the early nineteenth century. In 1988, roughly 5.6 billion tons of carbon were released as carbon dioxide from fossil fuel burning. Between 0.5 and 2.5 billion tons were released by deforestation and land-use changes.

- *Methane* (which makes up about 96 percent of "natural gas") is, molecule for molecule, a much more potent greenhouse gas than carbon dioxide. Each molecule of methane has 20 to 30 times the heat-trapping effect of a molecule of carbon dioxide. Each year, about 50 million more tons of methane enter the atmosphere than leave it.

Methane is produced when wood is burned inefficiently, when grasslands are set afire, and when fossil fuels are extracted and transported. Together, these sources contribute 100 to 200 million tons per year, according to EPA, and they account for recent atmospheric increases of this gas. But its biggest source (250 to 650 million tons per year) is a range of biological processes—among them, the rotting of organic matter in peat bogs, wetlands, rice paddies, landfills, and ocean sediments, as well as bacteria living in the entrails of farm animals and termites. Ice cores from Greenland and Antarctica show that the concentration of methane, which remained stable for 10,000 years at about 0.75 parts per million, has more than doubled since 1750. Today, it has reached 1.65 parts per million because sources are increasing and the atmosphere's ability to remove methane is being reduced by other pollutants. Methane's growth rate in the atmosphere is now about 1 percent per year.

- *Nitrous oxide* is produced in coal burning and forest fires, results from bacterial action on chemical fertilizers, and is a natural product of soil microbes' digestion. The concentration of nitrous oxide—also known as laughing gas—has gone up 10 percent since 1880, and by 0.2–0.3 percent annually in recent years. This gas stays in the atmosphere for centuries and eventually floats up into the stratosphere, where it also helps destroy the ozone layer.

- *Chlorofluorocarbons* (CFCs) are used in refrigeration and air conditioning, as blowing agents in packing materials and other plastic foams, and as solvents for cleaning such modern necessities as electronic parts; *halons* (closely related chemicals) are used in fire extinguishers only. Absent from the preindustrial atmosphere, they have both added tremendously to the warming problem in just the last few decades. CFCs are far more effective than carbon dioxide in trapping Earth's thermal radiation. Indeed, one molecule of the most dangerous CFC has about 20,000 times the heat-trapping power of a molecule of carbon dioxide! These high-power greenhouse gases also attack the ozone layer, and each CFC molecule can destroy 10,000 or more molecules of ozone. Concentrations of CFCs are now growing by about 5–7 percent annually.

Who first singled these gases out as environmental problems? The term *greenhouse effect* was coined by the nineteenth-century scientist Jean Fourier. His great insight was that the atmosphere possesses special properties that allow sunlight to enter but also allow absorption of radiant heat emitted from Earth's surface. Fourier first likened these qualities to the windows of a greenhouse. (Purists will note that *greenhouse* is a misnomer because such glass walls actually keep warm air from circulating and mixing with cooler air from outside besides absorbing radiant heat themselves. But the term accurately conveys the heat-trapping effect of this gaseous blanket.)

Later, in 1896, Swedish chemist Svante Arrhenius, who was acquainted with Fourier's findings, probed further. As a scientist living during the Industrial Revolution, he was concerned about what could happen if coal plants continued to spew carbon dioxide into Earth's atmosphere. He calculated that, if the concentration of carbon dioxide doubled in the atmosphere, the planet could warm by as much as 9 degrees Fahrenheit. Although this pioneering scientist relied on simpler assumptions than scientists use today, Arrhenius's pencil-and-paper calculations came astonishingly close to the best current estimate of 3–8 degrees Fahrenheit—an estimate reached by contemporary scientists with a far better understanding of weather forces and with access to much better data.

As this nutshell history makes clear, the greenhouse *effect* became

GREENHOUSE GAS PROFILE

Greenhouse Gas	Sources	Lifespan	Current Contribution to Global Warming
Carbon dioxide (CO_2)	Fossil fuels, deforestation, soil destruction	500 years	49%
Methane (CH_4)	Cattle, biomass, rice paddies, gas leaks, mining, termites	7–10 years	18%
Nitrous oxide (N_2O)	Fossil fuels, soil cultivation, deforestation	140–190 years	6%
Chlorofluorocarbons (CFC 11 and 12)	Refrigeration, air conditioning, aerosols, foam blowing, solvents	65–110 years	15%
Ozone	Photochemical processes	Hours to days in upper troposphere (1 hour in upper stratosphere)	12%*
			100%

*Includes small contributions from other minor trace gases.

the greenhouse *problem* only during the last century. Not until the Industrial Revolution did human activities begin contributing significant portions of carbon dioxide and other gases that trap heat in the atmosphere.

Although some measurements of carbon dioxide in the atmosphere had been made around the time of Arrhenius, not until 1938 was the body of facts great enough to support detailed scientific analysis. In that year, the British engineer G.D. Callendar compared the measured growth of carbon dioxide in the atmosphere and the long-term temperature records from 200 meteorological stations. Updating and confirming Arrhenius's now-famous calculation, he proposed a link between increased atmospheric carbon dioxide and global warming, which the prestigious British Royal Society greeted with skepticism. Undaunted, Callendar remained convinced that extra carbon dioxide was just what agriculture in the northern temperate zone needed and touted the atmospheric warming as a public good, a mighty defense against the "return of the deadly glaciers."

Although Callendar's data base was far superior to Arrhenius's, precise daily measurements of carbon dioxide in the atmosphere didn't begin until 1958, when Roger Revelle of California's Scripps Institution of Oceanography persuaded one of his graduate students, chemist Charles David Keeling, to measure carbon dioxide levels at a remote site, the Mauna Loa Observatory in Hawaii, to get an uncontaminated sample. Keeling began the best continuous record of carbon dioxide measurements on the planet, and the data gathered in Hawaii over the past three decades show that carbon dioxide levels have risen each year, from about 315 parts per million by volume of air in 1958 to about 350 today—the highest level in 130,000 years.

Very recently, other important evidence documenting the buildup of carbon dioxide was unearthed. In 1987, a team of French and Soviet scientists drilled an ice core 2,000 meters deep near a research station at Vostok, Antarctica. Analyzing air samples trapped in the ice, researchers estimated the atmospheric composition and local air temperature over the past 160,000 years. During a balmy interglacial period 130,000 years ago, carbon dioxide levels were just under 300 parts per million, dropping to 200 during the last Great Ice Age about 20,000 years ago. During the most recent 10,000 years, these levels climbed back to 280 as the planet entered the current interglacial warm spell. In the last 150

THE OZONE HOLE

Global warming was not the issue that first drew popular attention to the atmosphere's condition. It was the thinning of the ozone layer in Earth's upper atmosphere. This ozone loss is not fully understood, but scientists believe that chlorofluorocarbons (CFCs)—one of the main greenhouse gases—are the primary contributor.

The ozone layer, extending from about 18 to 30 miles above the earth, shields the planet from the sun's ultraviolet rays, which can harm many forms of life. This ozone is frequently referred to as "good" ozone, to be distinguished from the largely man-made "bad" ozone in the lower atmosphere that is the principal ingredient in urban smog. In human beings, damages from ozone depletion in the stratosphere can range from skin cancers to immune diseases to eye cataracts. Excessive ultraviolet radiation might also wipe out vast numbers of sea-going plankton—the vitally important microscopic plants and animals at the base of the marine food chain.

News about the ozone "hole" broke in 1985, when British scientists discovered a 40-percent drop in the springtime ozone levels high above Antarctica, and this finding was later confirmed by researchers reexamining NASA weather-satellite data. But speculation about possible links between CFCs and damage to the ozone layer traces back to 1974, when two U.S. scientists began wondering what would happen to the millions of pounds of CFCs produced every year as refrigerants, aerosol propellants, and foam-blowing agents. Since these "wonder chemicals" resist

(continued)

---

years, the level has reached 350, as recorded in Mauna Loa, and half of that increase occurred in just the last three decades. In short, we have changed the atmospheric concentration of carbon dioxide almost as much in a century as natural events did in the previous 10,000 years.

Ice cores can't tell us whether changes in atmospheric carbon dioxide levels cause temperature changes or follow them, but, as William Moomaw of World Resources Institute reports in *Orion* magazine, "What these measurements suggest is that the rate of change in temperature projected by the greenhouse theory is three to ten times faster than any of which we have knowledge—and that by 2030 we may well have brought about temperatures higher than any in 160 *millennia*."

THE OZONE HOLE   *(continued)*

chemical breakdown in the lower atmosphere, the scientists figured, CFCs would eventually drift skyward and collect in the stratospheric ozone layer starting some 15 miles above Earth's surface and extending upward. There, ultraviolet radiation would break the CFCs apart, releasing highly reactive chlorine ions to attack and destroy ozone molecules. The process is catalyzed by the ice clouds that form in the frigid air of the Antarctic winter.

Scientific expeditions to the Antarctic and other parts of the globe have borne this theory out. Using weather balloons and modified U-2 spy planes, scientists have measured an even greater ozone loss during the Antarctic spring and have also detected dangerous levels of chlorine in the frigid air. Other researchers have discovered similar, though less dramatic, ozone losses at the North Pole and over populated parts of the Northern Hemisphere. These discoveries have deepened scientific and public anxiety about the fate of Earth's ozone layer.

Responding to the early concerns, the U.S. government banned CFCs from most aerosol cans in 1978. But CFCs produced for other uses quickly filled the gap, and by 1987 nearly 2 billion pounds of CFCs were being produced worldwide each year. In the fall of 1987, 24 nations signed a historic treaty—the Montreal Protocol—to limit the production of harmful CFCs. Very recent research shows that a tougher treaty is now needed, but the Montreal agreement set a hopeful precedent for international negotiations on global environmental threats.

Although the 25-percent increase in atmospheric carbon dioxide since the beginning of the Industrial Revolution is bound to have transforming effects on our planet and the diverse creatures who call it home, it is important to understand that carbon dioxide is not a poison or even, at historical levels, a menace. Indeed, it is a vital part of Earth's chemistry, and it has shaped the evolution of life on the planet by stabilizing its temperature. Were it not for the atmospheric blanket of carbon dioxide and water vapor enveloping the planet, scientists estimate that Earth's temperature would be much lower, perhaps by as much as 60 degrees Fahrenheit. This natural greenhouse effect, as mentioned, created the conditions that allowed modern life to evolve.

Exactly how Earth has maintained its steady, habitable climate—unlike, say, Venus or Mars—still puzzles researchers. As National Center for Atmospheric Research climatologist Stephen Schneider and science writer Randi Londer note in *The Coevolution of Climate and Life*, many scientists have proposed theories on Earth's evolution, and most agree that some kind of greenhouse effect must have prevented Earth from turning back to ice.

HOW BIG IS THE PROBLEM?

For billions of years, Earth has moderated temperature swings by releasing back into space as much energy as it receives from the sun.

Read that last sentence again. And then consider that for the first time ever, during the last 100 years, human technology has proven it can alter the fundamental processes that govern the composition and "behavior" of the atmosphere. "A century of human industry has added significantly to the atmosphere's greenhouse-gas burden, and all indications are that this load will continue to increase indefinitely," writes Ben Patrusky in the National Science Foundation's journal, *Mosaic*.

As emissions of carbon dioxide and other gases have grown, more and more of Earth's radiant heat has been temporarily trapped and reemitted downward, warming the surface. Historically, carbon dioxide and water vapor increases contributed most to the enhanced greenhouse effect. In fact, since 1880, carbon dioxide has been responsible for two-thirds of the warming above background levels. But, during the last few decades, CFCs, methane, nitrous oxide, and other so-called trace gases have together contributed almost as much to the problem as carbon dioxide. The effect of these additional trace gases on global warming will be dramatic: they have accelerated the onset of the greenhouse problem by several decades. If higher carbon dioxide levels were the only problem, the world could be 3 to 9 degrees Fahrenheit warmer by 2075, according to the National Academy of Sciences. But, as if that weren't enough, the presence of the other trace gases pushes up the date by which the planet is committed to such a warming to about 2030.

As a sign of how quickly changes can overtake us, consider that an editorial on climate change in *Science* magazine in September 1977 warned, "By the year 2000 the carbon dioxide concentration will exceed preindustrial levels by about 25 percent." We reached that threshold already—in 1986, 14 years before the millennium!

In decades, the human race could induce changes that have been experienced in the past only over eons. In the last 10,000 years, the average temperature of the planet has increased less than 4 degrees Fahrenheit, but over the past 100 years alone, Earth's average temperature has risen by about 1.0 degree Fahrenheit. What's more, not all the effects of previous emissions have registered yet: no matter what we do from now on, according to Veerabhadran Ramanathan, atmospheric chemist at the University of Chicago, we must live with an additional 1.8 to 4.5 degrees Fahrenheit (1 to 2.5 degrees Celsius) warming *still to come* as a result of these same past emissions.

A warming of just 3.6 degrees Fahrenheit (2 degrees Celsius) will make Earth warmer than at any time in written human history. An increase of as much as 7.2 to 10 degrees Fahrenheit (4 to 5 degrees Celsius) will make our planet warmer than at any time in the last 1 million years.

As numerous studies of the greenhouse effect confirm, a 3.6-degree Fahrenheit (2 degrees Celsius) temperature change would bring with it profound and pervasive changes. Atmospheric scientists meeting in Villach, Austria, in 1985 projected some startling impacts worldwide should temperatures increase by 1.4 degrees Fahrenheit (0.8 degrees Celsius) per decade: extreme shifts in temperature in the high latitudes, more rain in the wet tropics, and a sea-level rise of as much as 1.5 meters by the middle of the next century—enough to erode beaches and coasts, destroy wetlands, and bring on severe flood damage to many low-lying countries.

The calculated temperature rise from greenhouse warming is only a global average, of course. High-latitude areas would experience temperature rises much greater than the global average and would face dramatic climate change if global temperatures rose 3–9 degrees Fahrenheit. Overall, polar regions would warm by two to three times the global average, while warming in the tropics would be limited to 50–75 percent of that average.

By bringing on in a matter of decades profound atmospheric changes that used to take thousands of years, we are also accelerating the rate of climate change. According to Stephen Schneider, the greenhouse

THE WORLD'S REFRIGERATORS

The polar regions—the world's refrigerators—are likely to experience the greatest warming of any places on Earth. They are also recognized as key levers in *controlling* the engine of climate around the globe.

"The Arctic and Antarctic are recognized to be vital components in shaping global climate and are likely to respond significantly to future climate change," write J. Barrie Maxwell and Leonard A. Barrie in *Ambio,* an international science journal. "Low temperature is the dominant feature of the climate of both the Arctic and Antarctic. It is the contrast between the cold conditions of these regions and the heat of the equatorial areas, combined with the rotation of the earth, which shapes the circulation patterns of the atmosphere—patterns which have a major controlling influence on global climate."

Antarctica is a particularly important "heat sink." Unlike the Arctic region, which is surrounded by populated continents and connected thermally to them, the continent of Antarctica lies latitudes apart, "a massive dome of ice (up to 4 kilometers thick)," these authors write, and "surrounded by ocean for over 20 degrees of latitude."

So climate and sea-level rise hinge to a great extent on Antarctica, the biggest glacier on Earth. At least 75 percent of the world's fresh water is contained within the Antarctic ice sheet, which represents some 160 feet of equivalent world ocean sea rise. "Any substantial changes in ice sheet mass will directly affect global sea level with worldwide consequences," write the authors of a 1988 paper for the International Geosphere-Biosphere Program.

Antarctica is also the key to past global climate, containing in its ice cores and ocean-floor sediments a detailed record of atmospheric chemistry reaching back hundreds of millennia. Returning from a trip there recently, Senator Albert Gore—a leading supporter of legislation to prevent global climate change—described in *The New Republic* a conversation with one of the scientists digging into the glacier: "High in the Transantarctic Mountains, a badly sunburned scientist talked about the ice core he and his team were pulling from a deep hole drilled into the glacier on which we stood. He interpreted the annual layers of ice and snow the way woods-

(continued)

THE WORLD'S REFRIGERATORS   *(continued)*

---

men read tree rings. 'Here is where the U.S. Congress passed the Clean Air Act,' he said, pointing to the beginning of the 1970s. Moving down the ice core, back to the early 1960s, he added, 'And here is where the world stopped atmospheric nuclear testing.' "

The earliest and most dramatic impacts of ozone depletion and global warming will be felt at the poles. Antarctica's frigid waters play as indispensable a role in the global carbon cycle as do the world's forests. Changes in the mass of the ice sheet could possibly throw off ocean circulation patterns as well as sea level. And changes in the mass of the ice, with its reflective whiteness, could also dramatically change Earth's surface reflectivity. If the temperature were to increase at high latitudes, more ice would melt, amplifying the local warming trend and possibly setting off a kind of chain reaction in which more ice melts and less heat is reflected back.

Such a big thaw in Antarctica could trigger several catastrophes. First, rapid breakup of the massive West Antarctic Ice Sheet could, over a couple centuries, raise sea levels 15 to 20 feet. Second, if solar reflectivity were altered at the poles by the disappearance of snow and ice, positive feedbacks that would reinforce the original effect could be expected, according to Maxwell and Barrie. Snow also slows heat loss, and it may well have a role in the formation of clouds, which themselves have a principal reflective role. Third, warming at either of the poles could melt permafrost, which some scientists fear could lead to sudden and large releases of methane trapped in frozen polar bogs. If all the methane trapped under the permafrost were released, some scientists argue, it would double the warming effect of all greenhouse gases released in the last century.

Finally, Antarctica is believed to play a particularly important role in the oceans' absorption and release of carbon dioxide. Scientists believe that the deep waters of the world's oceans contain 50 times the mass of carbon dioxide present in the atmosphere. These deep waters are "ventilated" near the continental slopes off Antarctica, so that any changes in circulation patterns could profoundly alter transfers of carbon dioxide between sea and air.

problem could change climate many times faster than natural climate changes have occurred in the past. For perspective, the natural rate has been less than 1 degree Fahrenheit per 1,000 years, but an artificially induced warming could amount to as much as 10 degrees Fahrenheit in less than a century.

Why is this matter of degrees so important? The Soviet scientist M. I. Budyko, who has long studied critical epochs in Earth's geological history, notes that rapid climate changes have usually been accompanied by massive biological changes—often mass extinctions. "There exists a simple ecological principle," he writes, "the more rapid the action of an unfavorable factor . . . the greater the damage caused to organisms." Although some types of organisms can stay one step ahead of climate changes by migrating long distances, many cannot. The EPA notes that trees, which reproduce by dropping seeds or cones, could not move poleward as rapidly as some scientists now predict that their climatic niche will move. By the same token, the microscopic plants that fix carbon in the upper warm layer of the ocean could not escape increased temperatures due to global warming or increased ultraviolet radiation due to ozone depletion. Entire species of these phytoplankton might perish before they could adapt.

Although such speculations lie well within the bounds of accepted science, no one knows *exactly* what the impacts of a global warming will be. And although scientists are working overtime to try to simulate possible effects and gauge their probability, we could well be in for some big surprises.

NO TWIN EARTH AVAILABLE

The first computer models used to predict climate pictured a simplistic world with only one continent, one season, and one cloud. Some of today's number-crunching machines complete as many as 500 billion arithmetic computations to try to simulate the weather in dynamic, three-dimensional terms. (For perspective, that involves about as many computations as humanity had collectively made up until 1940 or so.) But the climate's complexity still defies even the most sophisticated models.

The role that clouds play in Earth's weather is particularly mysterious, and it illustrates some of the problems that computer modelers face. Because cloud surfaces are white, they reflect nearly 30 percent of the sun's rays back into space, contributing to the planet's albedo, or reflectivity. But recent research suggests that clouds may also be part of a dimly understood climate-control system that includes the oceans and perhaps living creatures. One research team has proposed that even microscopic marine organisms may play a part in the weather by influencing clouds. U.S. and British scientists together found that in warmer weather oceanic plant plankton increase production of a natural byproduct called dimethylsulfide. This sulfurous gas reacts with oxygen in the atmosphere to form aerosols, tiny particles around which water vapor condenses to form clouds. The group hypothesizes that this plankton-cloud system might act as a global thermostat to help regulate Earth's temperature. When Earth warms— due to the greenhouse effect, for example—more clouds are formed and more of the sun's radiation is reflected away from the surface. The net result: a cooler Earth.

So do clouds warm or cool the atmosphere? Scientists aren't yet in agreement, and research has only begun to unravel the intricacies of Earth's climate. Other scientists from the United States, Europe, and Japan are also studying clouds and oceans—another major unknown force in climate change—looking for clues about the weather. One group, studying National Weather Service data for 45 U.S. cities, found that cloud cover above all but Fort Worth, Texas, has been steadily increasing since 1950. Air pollution? Airplane exhaust? The reasons behind such a buildup are unknown.

As Wallace Broecker writes in *Natural History* magazine, "Scientists struggle to increase our understanding of how the earth's environmental system operates in the hopes that we will be able to predict at least some of the coming consequences." But, as the "cloud problem" makes clear, scientists can't predict everything that could happen. Broecker warns, for instance, that scientific inquiries "have recently revealed a piece of disquieting information. Geological studies suggest that the earth's climate system resists change until pushed beyond some threshold; then it leaps into a new mode of operation." Likening the phenomenon to a radio dial that leaps to the next frequency point, he says that "the effects of greenhouse gas buildup may come in sudden jumps, rather than gradually." Such "nonlinear responses"—that is, responses disproportionately greater than changes in causes would suggest—could shock human societies and damage natural ecosystems even more than

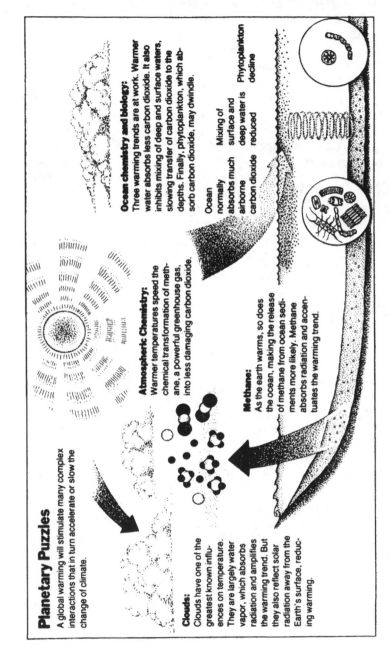

Planetary Puzzles

A global warming will stimulate many complex interactions that in turn accelerate or slow the change of climate.

Clouds:
Clouds have one of the greatest known influences on temperature. They are largely water vapor, which absorbs radiation and amplifies the warming trend. But they also reflect solar radiation away from the Earth's surface, reducing warming.

Atmospheric Chemistry:
Warmer temperatures speed the chemical transformation of methane, a powerful greenhouse gas, into less damaging carbon dioxide.

Methane:
As the earth warms, so does the ocean, making the release of methane from ocean sediments more likely. Methane absorbs radiation and accentuates the warming trend.

Ocean chemistry and biology:
Three warming trends are at work. Warmer water absorbs less carbon dioxide. It also inhibits mixing of deep and surface waters, slowing transfer of carbon dioxide to the depths. Finally, phytoplankton, which absorb carbon dioxide, may dwindle.

Ocean normally absorbs much airborne carbon dioxide

Mixing of surface and deep water is reduced

Phytoplankton decline

Source: New York Times, 27 January 1989

the gradual temperature increases and weather anomalies that most scientists predict.

The quintessential nonlinear response may be the seasonal ozone hole discovered over Antarctica in 1985. After the concentration of chlorine in the Antarctic stratosphere crossed a critical threshold (about 2 parts per billion), a new unanticipated type of rapid chemical reaction began, destroying ozone at unprecedented rates. This North America–sized hole—actually, a thinning of the gaseous shield that protects life on Earth from the sun's most damaging ultraviolet rays—is not a major factor in global warming. (The only connection is that the same CFCs that are destroying the ozone are also powerful greenhouse gases.) But its discovery proves that even after the initial surprise of an important "happening" in the stratosphere has worn off, scientists can't predict with confidence what is going to happen next. That's why Columbia University's Broecker has gone on record saying that Earth's in- habitants "are quietly conducting a gigantic experiment," the likes of which any responsible group of scientists and decision makers would "firmly reject."

Another wild card in predicting climate change is that there could be a kind of chain reaction of reinforcing effects. For example, as the polar regions warm, permafrost in the tundra could melt, releasing methane from compounds locked in the ground. This greenhouse gas would cause further warming, leading to even larger releases of methane into the atmosphere. Similarly, if the Arctic region warms, the ice covering the ocean there could melt. Thus bared, the uninsulated ocean would absorb more of the sun's direct energy than ice does and transfer heat to the atmosphere.

Responses like these are known as positive feedbacks because they tend to amplify the effect in question. Negative feedbacks, on the other hand, counteract or even wipe out a given effect. In many cases, the *net* effect is difficult for scientists to predict. As mentioned, it's hard to predict the effects of changes in cloud cover, which could tend to create either more solar reflectivity (because clouds are white) or more heat- trapping potential (because they absorb radiant heat). "The net result of the different [feedback] processes is a tripling of the warming caused by the doubled carbon dioxide levels," says David Rind of the Goddard Institute for Space Studies. "Yet, it is only the initial greenhouse effect due to the increased carbon dioxide increases and other trace gases that we know with great confidence."

With no twin Earth to experiment with, scientists employ "general circulation models" of the atmosphere, which are vast computer simulations of what the atmosphere is likely to do. These strictly numerical models are used for roughly calculating how the climate system responds to increases in gases. Most of the models divide the atmosphere into numerous cells or boxes—a kind of longitude/latitude/altitude grid. In each cell, the calculations take into account such variables as wind, temperature, moisture, incoming radiation, outgoing radiation, and the like. These models solve fundamental equations concerning the conservation of mass, energy, and momentum in each cell and then simulate exchanges among cells. But as complex as they are, the models still vastly oversimplify the workings of our planetary system. In such models, for instance, the oceans behave clumsily. Represented as a thin pond or a series of connected compartments, the oceans seem to behave more like flowing mud or dough than water carrying nutrients and heat in waves.

Penelope Boston, a researcher at the National Center for Atmospheric Research, sums up qualms about such oversimplification this way: "We run huge global climate models, in which we put in basic physical factors, and up to a certain point we can mimic the atmosphere, the latent heat, and movement of air masses. We have theories about certain parts of the behavior of oceans, and we have theories about certain parts of the behavior of the atmosphere. But we have no way of doing a computer model for the interactions *between* oceans and atmosphere. What I suspect is that all these subtle interactions elude our grasp."

Although scientists using these models can fairly accurately predict the degree of initial warming, given the presence of various greenhouse gases, they can't predict exactly how the rest of the system will react to the warming. Small changes in one of the planet's many systems—say, oceans, plant life, or atmosphere—can provoke enormous nonlinear responses in another. Scientists just don't know, as renowned environmentalists Barbara Ward and René Dubos put it two decades ago, whether the tiny pebble that gets moved is all that wedges a great boulder securely against the mountain wall.

Scientists also don't know exactly how climate change will affect various regions—a sizable problem because this is the level at which the changes will be felt and the level at which some of the initiatives needed to solve the problem must take place. For instance, even slight changes in the evaporation rate can alter humidity, cloud cover, and regional rainfall patterns, but models cannot reliably track these changes at the regional level.

Many scientists are concerned about relying too heavily upon limited, or even flawed, models. "My feeling is we overestimate our ability to predict," Wallace Broecker told the *New York Times* in January 1989. There "will be surprises, like the ozone hole over Antarctica. Therefore, we should be much more careful about what we are doing and much more observant about how the system works." Eli Katz of Lamont-Doherty Laboratory went even further. "The scientific goal at this point," he told William J. Cromie of the journal *Mosaic*, "is not to make predictions but to discover if predictability is possible on seasonal to decadal scales."

Although a major push is under way to understand feedback loops better, the fact is that most climate models today incorporate only a few of the major feedback processes. In 1989, the Environmental Protection Agency released a new study by Daniel A. Lashof and Dennis Tirpak, who assessed a range of feedback processes not now included in the climate models. Here are some of the mechanisms they found that could speed up global warming:

- As the temperature rises, methane from ocean sediments and tundra will be released faster.
- Vegetation that can't adjust rapidly enough to climate warming will die off, decaying and giving off more carbon dioxide. But this increase should be offset in part because carbon dioxide stimulates the growth of plants that absorb carbon from the air.
- The chemistry and circulation of the oceans could change as Earth warms, and oceans may no longer have as much capacity to absorb carbon dioxide.
- Higher temperatures could lead to higher levels of ground-level ozone (smog)—a heat absorber that could, in turn, accelerate global warming. (One study by Gary Whitten of SAI, Inc., predicts that smog over Nashville will worsen 40 percent with the temperature rise and stratospheric ozone depletion expected over the next 50 years.)

Scientists don't yet—and may never—have enough information and understanding to predict confidently the regional effects of global warming on natural systems, agriculture, and a host of other matters important to human beings. As Wallace Broecker warns, "We will know the results of the buildup of the gases [only] if our learning rate greatly accelerates." And if scientists don't yet know the full range of the direct effects and side effects of warming, who knows what other feedbacks remain to be discovered?

NO EXIT?

Considering how sensitive to stresses Earth's systems seem to be and how slow human beings have been to figure out how to live with that sensitivity, we might ask whether human experimentation on the planet ought, for ethical reasons, to be stopped—or, at the least, slowed down. Broecker took on this ponderous question in a 1986 article. To a great extent, he said, the greenhouse gas buildup has become "an inescapable by-product of our civilization." How, he asked, does one prevent carbon dioxide from escaping from power plants into the atmosphere? So far, we have no inexpensive and effective technology available to do so, or any program for reducing fossil fuel use. Several technologies have been proposed for "scrubbing" carbon dioxide from the exhaust gas produced by power plants. Unfortunately, all such approaches require considerable quantities of energy and add large amounts to the cost of power production. Even if they could be developed economically, disposing of the carbon dioxide by-product would still be daunting. So, Broecker concluded, "If five or so billion people are to be maintained on our planet, we must continue the greenhouse experiment. We are hooked."

But are we hooked? And, if so, how bad is our addiction? More and more, the question is: To what extent are greenhouse gases indeed essential to human survival?

Climate's Evolution
and the Industrial Revolution

And let them have dominion over the fish of
the sea, and over the fowl of the air, and over
the cattle, and over all the Earth . . .

GENESIS 1:26

For 200 years we've been conquering Nature.
Now we're beating it to death.

TOM McMILLAN
Former Canadian Minister of the Environment

Nature bats last.

BUMPER STICKER, 1989

Life before our species appeared goes back aways. For billions of years, cosmic forces shaped Earth, and land and air coevolved at an almost inconceivably slow pace to create a climate in which human beings and other creatures could flourish. Now, for the first time, humanity has the power to change the global climate. By releasing the huge amounts of carbon stored in fossil fuels over millions of years, we are speeding up and distorting the natural carbon cycle. We are intensifying the natural greenhouse effect—which is what makes human life possible—and turning it into a planetary menace. Thus, the German climatologist Wilfred Bach writes, "the carbon-dioxide problem becomes a central question for the co-existence of humans and the survival of mankind."

To put this concern into perspective, we need to examine both the great physical forces that drive climate and weather and the social and economic choices that drive us. In this interplay of environment and culture, we can begin to see what is within our control and what isn't and whether our actions can make a predictable difference.

25

A LONG VIEW OF WEATHER

Legend has it that after the Big Blizzard of 1888, an American told his European visitor, "Now, this is weather! What we usually have is climate." Weather is a fact of life. All human beings experience it from their first breath to their last. It has made a difference in battles, love affairs, and sports. But more than with most things, we take its daily variations for granted. It takes weather anomalies—blizzards, tornadoes, hurricanes, or temperature extremes—to remind us that the world is still governed by nature.

Each day's weather, of course, is just one frame in the constantly moving picture of a region's annual climate. As atmospheric scientist Stephen Schneider writes in *The Coevolution of Climate and Life*, it is "the instantaneous state of the atmosphere in space and time—a snapshot as it were." Conversely, climate is just a long view of the weather. It's based on long-term averages of temperatures, wind, rain, sunshine, and other elements of weather for specific areas. But climate is also what gives particular regions their unique flavor. It is vital to food production, supplies of fresh water, fisheries, and forests. A "favorable" climate influences where populations move and, ultimately, human health.

Ironically, until recently, humanity assumed that weather would always be unpredictable and that people could do little to avoid being buffeted by its natural vagaries. Indeed, many of our distant forebears took these vagaries as signs that the gods were angered or pleased. But with the spread of advanced communications, the development of meteorology, and, very recently, the use of weather satellites and supercomputers, we now have a much more informed view of weather and climate.

Basically, all weather begins with the sun. Located some 93 million miles from Earth and with a diameter 100 times greater, this celestial thermonuclear reactor showers our planet with energy as heat and light. What little is known about the early history of Earth's climate and the origin of life suggests that the planet was not always so warm and hospitable. But, during its 4.5 billion–year existence, Earth has developed a dynamic system for capturing the sun's light and warmth and maintaining an environment in which life could evolve and flourish.

The solar energy that reaches the outer edge of Earth's atmosphere averages about 1,130 watts per square yard. Each year, this shower of solar power adds up to 18,000 times more energy than humankind consumes as fuel and commercial energy. The sun, with surface temperatures approaching 11,000 degrees Fahrenheit (6,000 degrees Celsius), emits most of its energy as radiation—ultraviolet, visible, and infrared. Most of what reaches Earth is the visible "white" light that our eyes sense. The ozone layer between the troposphere and the stratosphere absorbs most of the ultraviolet rays, which can penetrate living tissue and alter genetic material, cause eye cataracts, and disrupt the immune system. What we feel as direct heat is infrared energy.

Almost a third of the sun's radiation that reaches the top of Earth's atmosphere is reflected back into space. Another 20 to 25 percent of the incoming solar radiation directly warms the atmosphere and clouds, which absorb the heat and then reradiate it in all directions. The remaining 45 percent or so reaches Earth's surface either directly or as diffuse radiation that the clouds and atmosphere have scattered. Although our atmosphere is relatively transparent to shortwave radiation from the sun, it also absorbs some infrared radiation emitted by Earth. Throughout this process, color makes all the difference. White surfaces—clouds, ice, and snow—tend to reflect sunlight, and the darker oceans and land masses absorb it.

More than half the solar energy that reaches Earth evaporates water or melts ice and snow. Some sun-warmed ice evaporates directly into water vapor that carries the sun's energy long distances before freezing or condensing and returning the heat to the environment. Climatologists call this phenomenon "latent heating" because the heat is merely stored in the new vaporized state until it is released in condensation. The sun's energy also directly warms air and land surfaces. This process is called "sensible heating." Such heat is transported, for instance, when warm, dry winds blow into cooler regions or when bodies of water warm adjacent land masses. In this way, oceans and lakes buffer the climates of coastal regions, keeping temperatures more moderate than those found inland.

In a nutshell, weather issues from the sun. The tropics are warm because more intense solar radiation falls in equatorial zones. Polar regions are cold because they receive less solar energy. The temperature differences between the equatorial and polar regions drive the wind and water cycles that determine global weather patterns.

However mechanistic this description may seem, climate has a long and fascinating history. Life before humans consisted of a complex series of slow climatic shifts. During most of that time, the planet appears to have been generally inhospitable to life. During the seven relatively brief glacial epochs, the longest of which lasted a mere 18 million years, Earth was "normally" colder than now. The climate we have known during the past several thousand years has been unusually benign, when viewed from the perspective of Earth's 4.5-billion–year history. The only significant recent exception is the Little Ice Age of the Middle Ages, when temperatures were 5 degrees Fahrenheit lower than they are today.

Crispin Tickell in his book *Climate Change and World Affairs* gives us a sense of just how unusual today's climatic conditions are. "We live in a tiny, damp, curved space at a pleasantly warm moment." If Earth's history were compacted into 46 years—about one year per 100 million—Tickell writes, "then the dinosaurs died just over 6 months ago, the present human breed emerged about a week ago, our counting system before and after Christ began less than a quarter of an hour ago, and the Industrial Revolution has lasted just over a minute." Looking at the present interglacial period on this 46-year time scale, the major ice ages would have occurred 9 ½ years ago, 7 ¾ years ago, 6 ¼ years ago, 4 ½ years ago, and about 3 years ago. Given the perspective of past centuries and insights from the new science of paleoclimatology, human beings have been fortunate. "Man has made his appearance on Earth at a time of unusual climatic conditions," according to a report of the International Institute for Applied Systems Analysis.

Climate does change. But changes over millennia or even millions of years are one thing, and changes over a few decades or a lifetime are quite another.

THE HUMAN HAND ON THE GLOBAL THERMOSTAT

What are human beings doing that affects climate? And how does human influence on climate compare with nature's own fluctuations? These questions can't be answered simply. Of all the aspects of Earth's environment, climate remains one of the least understood because so many natural variables affect it and keeping track of them is a daunting task. "It is hardly surprising that the variability of climate is a recent discovery of science," writes scientist and futurist John Gribbon. "It was only in 1840 that official records of temperature, rainfall and so on began to be kept at the Royal Greenwich Observatory in London."

In the nineteenth century, debates raged over how radical earlier climate changes had been and to what extent they could reoccur. To most, it seemed shocking that large parts of North America and Europe could be covered with ice. But the belief that climate over human history had been fairly stable was undermined, writes Gribbon, as more facts from the geological record emerged. During the last century, a comparative wealth of information about past climatic history has been uncovered in sediment and ice cores, tree rings, and fossilized pollen and vegetation. Ironically, at the same time that we are learning more about how the climate system works, human activities are artificially inducing climate change.

Perhaps the least understood determinant of Earth's climate is the human element. Until roughly the last 100 years, no human event or activity has been able to change Earth's climate radically. But that humankind has altered regional and local climates to some extent over historical time is beyond question. Relative newcomers in an age-old process, human beings have upset regional and local climate again and again.

Historically, by cutting down forests, overgrazing rangelands, or failing to let agricultural lands lie idle long enough to replenish their fertility, humankind has created deserts and wastelands. As Rutgers biologist David Ehrenfeld writes of so-called desertification, "When the vegetation is reduced, more light-colored bare sandy soil is exposed, and this increases the albedo or reflectivity of the landscape. When the albedo increases, more sunlight is reflected, and the land becomes somewhat cooler. Air passing over this landscape is heated less than usual and tends to rise less. This, in turn, decreases cloud formation, which decreases rainfall. Lower rainfall prevents the regrowth of vegetation, the albedo increases further, etc. And so the deserts expand."

Processes like those Ehrenfeld describes can take millennia or, more recently, mere decades, and can utterly transform the landscape. Lebanon's renowned forests have steadily given way to surrounding desert under human pressure. The gradual transformation began when the biblical King David ordered trees cut to build ships, some 3,000 years ago, and continued until World War II, when British forces stripped firs and oaks from the northwestern slope of Mount Lebanon to construct a railroad between Tripoli and Haifa. The wholesale deforestation beginning about A.D. 600 around the ancient sacred city of Teotihuacan in central Mexico took roughly 200 years to occur, destroying the region's soils and civilization. Unsound farming practices and a severe drought in the U.S. Midwest in the early decades of this century combined to pave the way for the Dust Bowl of the 1930s and the human suffering depicted in *The Grapes of Wrath*. In Indonesia and elsewhere in Southeast Asia, efforts to harvest the remaining virgin hardwoods have led to devastating floods and soil erosion. Other examples abound.

The question of how land use influences the rise and fall of nations has long engaged historians. But land-use decisions have also led to changes in surface temperatures, rainfall, and other local conditions of concern to climatologists. Phillip Fernside and others studying the Brazilian experience have argued that deforestation changes the Amazon basin's ability to recycle moisture. Because the Amazonian forest transpires about as much moisture each year as the Amazon River carries, it is feared that if people continue to deforest land and convert it to pasture, regional rainfall will drop so dramatically that both the forest and the farmland will be doomed. Many experts also argue that land use has had incremental effects on global climate as well.

Another way that human beings have affected climate locally is through urbanization. Most cities of the world are capped by masses of warm air, like a dome of glass, that rises up to a height of about 400 feet. It is common to find differences up to 11 degrees Fahrenheit between cities and surrounding rural areas, with the biggest differences showing up at night. In the summer in New York City, for example, temperatures are routinely 10 to 15 degrees higher in the inner city than in neighboring Westchester County to the north. These "heat domes" remain over the cities, and often it is only when wind speeds reach 15 miles per hour that the warming effect can be shaken.

Car exhaust, industrial smoke, waste heat, tall buildings that hold heat and block winds, and heat-absorbing pavement, such as malls, parking lots, and roads that replace yards, fields, and woods, all contribute to this effect. Smog, too, can create a blanket over cities that traps heat, thus creating a kind of mini-greenhouse effect. Collectively, these urban heat domes are not large enough to raise the planet's temperature. As William W. Kellogg and Robert Schware write in *Climate Change and Society*, this human-generated heat currently accounts for only about 10 terawatts (10 million million watts)—1 ten-thousandth of the sun's heat absorbed by Earth's surface—and thermal pollution would have to increase 100 times to account for even a 1.8-degree Fahrenheit change of average surface temperatures.

Carbon dioxide buildup, however, is a different story. No other human factor so influences climate change on such a wide scale. Particulates (such as soot) and aerosols released through normal human activities can affect regional weather but not global climate. Urban growth and paved roads may alter the local albedo. But only the greenhouse effect and nuclear explosions are large enough to reset the global thermostat.

Twenty years ago, write Kellogg and Schware, "the prevailing attitude was that the global climate system was probably too enormous and too stable—some have referred to it as 'very robust'—to respond to anything we could do to it." But that attitude changed rather suddenly in 1971 when a landmark international meeting of scientists studying human impacts on climate found participants concurring that climate change was a very real and near-term possibility.

So far, there is no scientific consensus on when we should be able to detect an "anthropogenic signal"—a sign that *human* activity has begun to change Earth's climate. But many scientists expect that signal to come in a matter of years. In 1988, James Hansen of the Goddard Institute for Space Studies told Congress that human-made increases in temperature may have already been detected. And atmospheric chemist David Rind of Goddard states that the carbon dioxide and other greenhouse gases now in the atmosphere have *already* trapped enough heat to make the earth warmer than it has been anytime in the last 20,000 years.

The timing of this signal—what William Moomaw of World Resources Institute calls the "human fingerprint" on climate—remains of concern to scientists, especially those working on short-term responses to the greenhouse problem. But the far more important issue is how long we will have to live with a global warming once it arrives. As William Kellogg writes, "the natural sources and sinks of carbon dioxide must have been in near-equilibrium" for millennia before the Industrial

Revolution began. And natural control mechanisms that reestablish the atmospheric equilibrium work glacially in comparison to human life-spans, so that we can expect it to be centuries before human disturbance of the atmospheric carbon dioxide content will return to its original state of equilibrium once it has been disturbed.

These scientists' pronouncements on the long-lived effects of med-dling with Earth's atmosphere make efforts to reduce human influence on climate change all the more urgent. With ripple effects lasting some 20,000 years—not to mention the devastating immediate impacts already discussed—how we live today is a matter of enormous conse-quence.

A FUELS PARADISE

The causes of human-induced climate change are varied, numerous, and interactive. But the chief cause is 200 years of ever-increasing fossil fuel use. In a way, climate change is the bill for the Industrial Revolu-tion coming due.

The Industrial Revolution began in England just before 1800. Facing a timber shortage, England began burning coal and peat to operate steam engines. During the last century, the use of oil, coal, and natural gas has provided the basis for the world's industrial expansion and all the economic and social benefits that came with it. The high-energy life style, now the hallmark of Western industrialized countries and of the elite throughout the world, began in the United States with the drilling of the first oil well at the time of the Civil War. As energy analyst Wilson Clark details in *Energy for Survival: A New Era*, "It was here that oil was first drilled and exploited on a large scale. It was here that coal and steam condensed a continent into a few days' train journey. It was here that the electric light and the electric distribution system were invented. It was here that abundant energy and mass production made the cheap automobile ubiquitous . . ."

Until around 1850, the United States ran largely on horsepower and such renewable resources as wind, water, and firewood. By 1885, however, coal had surpassed wood as the main fuel source, much of it used for railroads and the steel industry. Supplying about two-thirds of the nation's fuel, coal had become king and would remain so well into the next century.

Oil did not really enter the U.S. economy on a large scale until 1860, when 500,000 barrels of crude were produced—just one year after the first modern drilling rig was erected near Titusville, Pennsylvania. The United States was not the first society to exploit petroleum, however. Among other applications, it had been distilled for lamp fuel as early as 400 B.C. near the Ionian Sea. "But it was only in the Western industrial world of the 1850s," writes Clark, "that the confluence of science, technology and society were at the juncture necessary for oil to be seized upon as the concentrated fuel that would replace [first whale oil and then] coal."

The next major milestone for fossil fuels was 1877, the year Nikolaus Otto invented a practical gasoline-fueled engine. The forerunner of all internal combustion engines used today, this engine was first commercially adapted by Ford and Oldsmobile around 1900. In 1879 came the invention of Thomas Edison's electric light, and in 1882, the first electric power generating and distribution system in New York City. Edison himself described this marvel as "the biggest and most responsible thing I had ever undertaken."

It was also around 1900 that electrical appliances began emerging in the United States, with the introduction of everything from hot plates to radiant-panel heaters, coffee grinders to electric cigar lighters. And in 1901, the famous Spindletop gusher in southeast Texas seemed to assure Americans that they wouldn't ever again have to worry about how to power the new high-energy society of which these gadgets were a part. As Denis Hayes writes in *Rays of Hope: The Transition to a Post-Petroleum World*, "heady success overpowered prudence" after the find, and "the inevitability of oil exhaustion became an abstraction—hard to grasp and comfortably remote—as huge discoveries were made in Oklahoma, Louisiana, California and Alaska."

Electric power capacity grew steadily throughout the first half of the twentieth century, increasing, according to Clark, by about 1 to 2 percent each year until World War I. The pace of electrification accelerated (fueled increasingly by coal) after the war, and growth averaged 7 to 9 percent a year from the mid–1930s until World War II began.

By 1940, the steady increase in coal consumption was slowing while crude oil consumption more than tripled. From 1940 to 1960, Clark explains, "railroads switched almost completely from coal to diesel power; an entire jet fleet of passenger airplanes developed, the auto population experienced phenomenal growth. Air conditioning, television, central heating, and clothes washers and dryers made the transition from luxuries to necessities; a mechanized agricultural industry began using 16 billion gallons of crude oil a year; and monstrous glass towers inspired by Lever House and the United Nations building came not only to dominate the skylines of the [great] cities, but to sprout up even in the small cities of America."

Globally, the rise in energy consumption has also been dramatic. In 1950, the world consumed 80 "quads" (quadrillion British thermal units) of energy. (Roughly speaking, a BTU is about as much energy as that emitted by one match when burned.) In 1988, the United States alone consumed 80 quads and the world used four times that much. Only after World War II did the extent of global dependence on fossil fuels become clear. Energy crises, calculated or actual shortages of fuel, were the result of this dependence. Two major events set the stage for the first energy crisis. On October 17, 1973—a date that S. David Freeman, former chairman of the Tennessee Valley Authority, called "energy Pearl Harbor Day"—a few major oil-producing nations announced that they would halt deliveries to the United States and The Netherlands. Then in 1974, writes Freeman, "the oil-producing nations dropped the other shoe, so to speak, by more than doubling the price of crude oil."

The Arab oil embargo and OPEC's price hikes of the 1970s exposed the vulnerability of the United States and other industrialized nations' petroleum-based economies. As Colin Norman writes in *The God That Limps*, "the 1973–74 oil embargo and the associated rise in oil prices forcefully demonstrated the deep structural changes that had taken place in the world oil market over the previous decade. Between 1950 and 1973, world oil production increased from 4 billion to 20 billion barrels per year, climbing at a steady 7 percent annual rate. This gusher of cheap oil—its price declined substantially in real terms in this period— was poured into the growing automobile fleets, residential and industrial boilers, chemical factories, and electricity-generating plants of the industrial world. Energy-intensive technologies developed in this period helped change the shape of cities, factories and daily lives. But these developments carried a hidden price: growing dependence on oil supplies halfway around the world."

By the early 1970s, when the oil price shocks registered, there was no question: we were hooked. As Norman writes, "the realization had finally begun to sink in that industrial technology had become danger-ously dependent on a source of energy that would eventually run out and that the patterns of production and consumption established during the fifties and sixties would be unsustainable over the long term." Accord-ing to World Resources Institute research, as a result of the sharp rise in prices after the first oil crisis, the industrialized world paid $1.5 trillion (1984 dollars) more for oil imports between 1973 and 1981 than it would have had oil prices remained at the 1972 level.

The second oil price shock came in 1979, when Iranian production was sharply curtailed. At the time, writes Norman, OPEC members took advantage of a shortfall in oil, boosting prices from $13.77 to $28.45 per barrel between January 1979 and January 1980. "In a decade, the price of oil had jumped by a factor of 15," he writes. "Cheap energy, which had exerted a major influence on technological change for a generation, had finally passed into history."

As painful as the energy crises of the 1970s were, however, they led to a deeper understanding of how energy can be better used. After the oil shocks of the 1970s, world demand for oil dropped. World oil production fell from 62,360,000 barrels per day at its height in 1979 to 55,690,000 in 1987, according to Energy Information Administration statistics.

Despite this decline in oil production, total world consumption of fossil fuels in general is steadily rising. According to the *British Petro-leum Statistical Review of World Energy*, world use of commercially traded fuels went up incrementally during the 1980s, 1 percent between 1982 and 1983, 3.7 percent between 1983 and 1984, and 2.6 percent between 1984 and 1985. Oil use fell, but coal use rose enough to raise total fossil fuel energy consumption again in the 1980s. Global coal consumption, which filled the gap left by costlier oil in some countries, went up by 30 percent between 1977 and 1987.

Fortunately, energy-efficiency improvements around the globe did slow the growth of energy use globally. As José Goldemberg and his coauthors write in *Energy for a Sustainable World*, some of those deciding on energy use around the world are beginning to turn away "from the historical preoccupation with expanding energy supplies to examine how energy can be used more effectively in providing such services as cooking, lighting, space heating and cooling, refrigeration, and motive power." These forward-looking decision makers are finding

Sources of Energy
Percent of total U.S. energy use supplied by each
type of power. In 1987, the United States
consumed a total of 76.3 quadrillion BTU's of
energy.

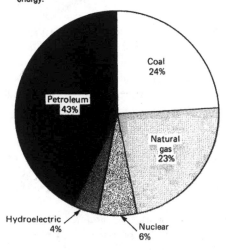

ways to fuel the economy with far less energy than previously thought
necessary, "and as a result the historically close correlation between the
level of energy use and economic well-being has been broken."

To prove this de-coupling of energy and economic growth, Goldemberg and his coauthors show that between 1973 and 1985, total energy
use per capita in the Organization for Economic Cooperation and
Development (OECD) countries—the world's most industrialized market economies—fell 6 percent while the per capita gross domestic
product increased 21 percent. Specific countries showed even more
impressive advances. In the United States, they write, per capita energy
use fell 12 percent while per capita gross national product rose 17
percent. In Japan, per capita energy use fell 6 percent while per capita
gross national product rose 46 percent—one powerful reason that its
economy over this period has been the world's envy.

As ways are found to use energy more efficiently, these gains can
only grow. In the United States, there are hopeful trends in fossil fuel
consumption. Between 1973 and 1987, U.S. industry cut direct burning

of fossil fuels by a fourth as it increased its use of electricity. During those same years, residential and commercial buildings in the United States cut their direct fossil fuel use by 21 percent. During the 1980s, however, energy use in both transportation and electric power generation was on the rise. Because increasing use in some sectors is partially offsetting declines in others, this is a "good news, bad news" story.

FLYING HIGH, DRIVING MIGHTY

Energy use in U.S. transportation has grown explosively since mid-century. Air passenger travel and air freight, by far the least energy-efficient ways of moving people and goods, have skyrocketed. Automobile use has also grown phenomenally. Between 1950 and the mid-1970s, for example, automobile traffic increased 142 percent, while the amount of energy needed to power vehicles increased 171 percent. Globally, the vehicle fleet has grown from 53 million in 1950 (75 percent of them in the United States) to 386 million in 1986 (65 percent of them outside the United States). That increase translates into 5.7 percent annual growth in the automobile population—far higher than annual growth in the human population.

Invented as a mechanical replacement for the horse and buggy, the automobile gave us unprecedented freedom of movement and set in motion an age of high-speed travel. But cars like today's gasoline guzzlers have their detractors. Ernest Callenbach writes in *Ecotopia* that "the car, delightful as it might have been in a nineteenth century land of wide-open spaces, is really not appropriate for urban living patterns. It is, in fact, in that class of items known as 'positional goods': things that are undeniably good to have, but only so long as only a few other people have them too; if they're too common, they interfere with everybody's enjoyment."

The historical trend in the United States for driving—cars, trucks, buses, and motorcycles—is up, up, up. In 1900, there were some 4,200 automobiles being driven in the United States—most of them steam driven or electrically powered, and only one-quarter driven by internal combustion, according to Wilson Clark. Since then, Americans haven't increased the number of miles they travel per car, but with more and more people owning more and more vehicles, the total number of vehicle miles traveled by the fleet has quadrupled since 1950, increasing steadily at an average rate of 3.5 percent each year for the last 40 years, according to figures from the Federal Highway Administration. As a result, U.S. fuel consumption for ground transportation has increased more than sixfold during approximately this same period, going from 20 billion gallons per year in 1940 to 135 billion gallons in 1988.

Regardless of what Detroit and the other auto capitals of the world might think, we've got car trouble. Burning a single tank of gasoline produces 300 to 400 pounds of carbon dioxide, a principal greenhouse gas. In the United States, motor vehicles are responsible for about 31 percent of all carbon dioxide emissions. (Worldwide, the figure is 15 percent.) Cars, trucks, and buses also use and emit roughly 1 of every 8 pounds of chlorofluorocarbons (CFCs) manufactured in this country. And, for all Western industrialized countries, cars are the main sources of carbon monoxide, hydrocarbons, and nitrogen oxides—other compounds implicated in climate change. In the United States, almost three-fourths of all carbon monoxide emissions come from vehicles.

How much damage our transportation system will do to the atmosphere depends on how many vehicles are on the road and how much of these offensive pollutants they emit. And emission rates of carbon dioxide, in turn, depend on *how efficiently this fleet burns fuel*. Unfortunately, no real fuel economy gains have been made in the U.S. fleet since 1982. In fact, the average fuel efficiency of new cars in this country slipped slightly in 1988 to 25.8 miles per gallon, from 25.9 the year before. As transportation analyst Michael Walsh puts it, "The fuel efficiency wars of the late 1970s have been replaced by the horsepower wars of the 1980s."

Even though Asian-built cars are far more fuel efficient than U.S. vehicles and are giving cars built here and in Europe a run for the money, there are no net gains in fuel efficiency worldwide either. Television ads drive the point home: in the years since the oil crisis, international marketing highlights auto performance and styling, not efficiency.

Obviously, even if fuel efficiency comes back into vogue, as some oil-price watchers predict it will, fuel savings—and emissions savings—will be canceled by sheer growth in the number of cars unless efficiency gains are truly spectacular. But the prospects for engineering miracles and less crowded highways don't look good. Growth in the Western industrialized countries has slowed as comparatively stable and prosperous populations reach a saturation point. In overcrowded Japan, for instance, you can't even buy a car until you prove that you have a place to park it. But in many parts of the world, car hunger won't be sated until the kind of gridlock that routinely paralyzes such cities as Cairo and Los Angeles makes owning a car more trouble than it's worth. In the developing countries especially, rapid population growth and urbanization are, literally, driving forces. Asia's population is expected to double (from its 1960 level) by the year 2000, and Latin America's to rise by 150 percent. The 1985 world population will almost double by 2025, and the auto population can be expected to rise along with it, traffic and OPEC permitting.

Air travel is also a growing consumer of fossil fuels, mostly because more tourists are traveling by plane than ever before. In 1984, the most recent year for which figures are available, 7.9 percent of our transportation fuel was consumed by jet planes. Like automobiles, airplanes have become more efficient. In 1984, they consumed only half as much fuel per passenger mile as they did in 1970—a huge reduction in energy waste. Nevertheless, with more planes flying, more jet fuel is burned. In 1984, total consumption of jet fuel was 20 percent higher than in 1970. That total translates into about 50 million tons of carbon dioxide for the year.

INDUSTRY'S ENGINES

The importance of the energy industries to environmental problems goes far beyond the direct impacts of energy supply. Think for a moment about how dependent *all* industries are on oil, gas, coal, and other fuels. It's not just that fossil fuel extraction and refining influence the economy and pollute the environment, but also that all of the rest of our industrial production depends on these industries and consumes a substantial fraction of what they produce. In any case, it's clear that the way industry uses energy in the future will affect our national wealth and the risks we face from rapid climate change and other major environmental problems.

About one-third of the nation's current energy use—and more than a quarter of all carbon dioxide pollution—is by major industries. Industry also emits a larger fraction of the other important pollutants as well. In 1986, industry accounted for 13 percent of emissions of the oxides of sulfur (78 percent if electric power production is included) and about 20 percent of the emissions of oxides of nitrogen (exclusive of power generation) that contribute to the risk of acid rain. Industrial activity also led to the release of 39 percent of the volatile organic compounds that cause urban smog and 19 percent of all carbon monoxide emissions.

Throughout the postwar period, industrial consumption of fossil fuel energy rose steadily as industrial production expanded. Wilson Clark notes that "the nation's prime movers—the various engines that perform the work of American society—grew from a total of 2.8 billion horsepower in 1940 to 21 billion horsepower in 1971." The rate of industrial and technological growth, he points out, so exceeded the growth in American population that "while the number of people increased by 54 percent during these three decades, the prime movers of high-technology increased by 750 percent."

In the last 25 years, the United States' rate of growth in both population and industrial energy use has slowed, but the totals continue to increase. Population has increased by less than 10 percent since 1963, while industrial energy use has gone up over 20 percent.

During this period, the mix of energy sources used by industry has also changed, as electricity replaced direct heat. Indeed, electricity use in industry doubled between 1963 and 1988, rising from 7 to 14 percent, while direct coal-burning fell from about 25 percent to 13 percent.

In 1987, industry accounted for a little more than one-third of total energy use, and almost two-thirds of this amount was consumed directly as fossil fuels for heat and feedstocks. But not all industries consume equal shares of the total. The vast majority of industrial and commercial businesses, according to S. David Freeman, are not heavy users of energy. "Their energy use resembles residential consumption," he writes: "fuel for heating and cooling and electricity for running machines." But a few energy-hungry industries consume far more. Just five industries—iron and steel production, nonferrous metals manufacturing, paper and pulp products, chemicals and feedstocks, and stone, clay, and glass production—consume about three-fourths of all the energy used in manufacturing. According to the Energy Information Administration, fully a fifth of all manufacturing energy is consumed by the chemicals industry alone. As Wilson Clark points out, "The fossil fuel resources—coal, oil, natural gas—that are used for the production of power in our society have increasingly been diverted to the manufacture of synthetic materials."

As energy prices rose over the last decade and some developing countries with newer factories began producing such energy-intensive goods as steel and aluminum for less than the United States could, the U.S. industrial demand for primary energy fell. By 1987, industrial energy demand was more than 10 percent below its 1977 levels. But since then, consumption has risen again—mainly because production of such durable goods as tools and appliances is on another upswing.

Although industrial energy use is difficult to predict, it will certainly continue to represent a significant fraction of total energy demand, both in the United States and worldwide, for the next century. The challenge ahead is to squeeze more output from each ton of raw material and to reduce the rate at which harmful pollutants are released in the process. Shifting heavy manufacturing from the United States and Western Europe to the Third World merely moves the sources of greenhouse gases (and the "precursors" of acid rain) unless the most efficient and least polluting technologies are also shifted overseas.

AN ENERGY-HUNGRY AGRICULTURE

Modern agriculture is a major—and growing—contributor to the greenhouse problem. Over the last four decades, as agriculture has shifted from family farming to large-scale corporate farms, it has become increasingly industrialized. More and more fuel is needed to operate farm machinery, and the machines themselves require more energy than the smaller, simpler devices of bygone agricultural eras. Agriculture also uses chemical fertilizers and fossil fuel-based insecticides and herbicides that require additional energy to produce. Worldwide, agricultural practices currently contribute 14 percent to global warming. As developing countries mechanize their agriculture and increase their use of synthetic fertilizers, this percentage could rise.

A major agricultural contribution to the greenhouse effect is nitrous oxide emissions from bacterial action on the chemical fertilizers used in intensive agriculture—in the industrialized countries and in the "green revolution" agriculture in the Third World. This gas is given off when soil microbes digest chemical fertilizers; the more inorganic, nitrogen-based fertilizer farmers apply, the more nitrous oxide goes into the air. This potent pollutant also helps destroy the ozone layer.

Another greenhouse gas produced both in human activities and in other natural processes is methane, an inevitable by-product of livestock raising and food production. About a third of the methane in the atmosphere arises from the incomplete combustion of fossil fuels and from coal mines and natural gas fields. Bacterial action generates another large fraction of global emissions. Major agricultural sources are flooded rice paddies and wetlands, ruminant animals, and termites. (Termites, which produce about 8 percent of all methane, are incredibly numerous. According to John Firor of the National Center for Atmospheric Research, for every person on the planet there are about 1,500 pounds of termites.) Other human activities, including forest burning and garbage burial in landfills, also contribute to methane emissions. Methane concentrations have been increasing by about 1 percent a year. Normally, methane is removed from the atmosphere naturally, but carbon monoxide, a by-product of vehicle exhaust, reduces this natural cleansing capability.

BUILDINGS—EMBODIED ENERGY

Most of our architecture fights with climate, and the greenhouse problem is now in part a buildings problem because so much fossil fuel energy is used to heat and cool poorly designed buildings. Today, roughly one-third of all the energy consumed in the United States is used in commercial and residential buildings. U.S. industry uses another third of the total for process heat as well as for space heating and cooling.

Although ingenious, twentieth-century heating and cooling technology is predicated on cheap, ready supplies of fossil fuels. Somehow over the centuries, we *un*learned the lessons of the ancients in designing with natural forces in mind. For thousands of years, the Greeks, Romans, Chinese, Native Americans, and others located and built their homes and communal buildings with wind, water, and sun in mind. In the Roman Empire, "solar access rights" were written into law when fuel wood shortages put a premium on solar heating, and well-placed windows were a status symbol. In ancient China, shamans were enlisted to find the most naturally auspicious spots for homes (and graves). Sages from Socrates to Pliny the Younger prized carefully oriented homes that trapped heat in winter and capitalized on cool summer breezes.

Americans are different in part because they descended from one of the few groups of colonists who never adopted native architecture. Even the simple "long houses" of Native Americans were more effective defenses against the harsh New England winters than were the transplanted European-style houses that the early colonists built.

This die-hard defiance of nature, part of the much-admired frontier spirit, has backfired. As S. David Freeman writes in *Energy: The New Era*, "Americans expect—indeed, they insist upon—sufficient energy to maintain the temperature they desire in every inch of indoor living and working space no matter how hot or cold it may be outdoors . . . the typical American building is now too hot in the winter and too cold in the summer." And the economic and pollution costs of maintaining such temperature extremes are themselves extreme.

Houses built between 1940 and 1975—just over half the standing housing stock today—are the most wasteful of energy. According to housing expert Fred Dubin, the tract houses that sprang up from Levittown to Los Angeles during this era are especially flagrant examples of energy-inefficient housing. "They were built cheaply and quickly with poor insulation," says Dubin. "Builders put up single-glazed windows and eliminated shutters."

The modern skyscraper also exemplifies unnatural architecture. A sheer block of glass and steel, it traps so much heat that prodigious amounts of energy are needed to air condition its interior—often, even in winter. According to one estimate, the World Trade Center in New York City consumes more energy than the entire city of Rochester in upstate New York.

In the last decade, we Americans have become more aware of how energy wasteful our homes and workplaces are. According to the New York State Energy Office, even the World Trade Center has been subjected to an energy retrofit. But homes and buildings have a long way to go.

Lee Schipper, an energy expert at Lawrence Berkeley Laboratory in California, claims that builders have plenty of new energy-saving technologies available to update architecture. But "what we're up against is an institutional bias against investing in conservation." Most consumers, says Schipper, won't go for energy-efficient extras because they don't improve the houseowner's status: "You can't tell from the outside how thick the walls are." And builders take their cues from buyers. According to Michael Bell, a former National Association of

Homebuilders official and now a builder, they are "building for what the marketplace will stand" and won't add conservation features that will raise costs consumers—be they homebuyers, investors in commercial property, or landlords—don't want to pay.

The United States falls behind other countries when it comes to the energy efficiency of its building stock, say energy experts. In contrast, the Swedes build houses to the world's highest standards of efficiency; theirs are at least twice as energy efficient as ours. Japanese housing stock is also way ahead, says Paul Kando of the Center for the House in Washington, D.C. "We don't have an energy policy in this country, nor do we have a housing policy for that matter," says Kando. "If you treat energy as an extraneous thing in building, as a kind of afterthought, you make a big mistake."

Sweden dedicated itself to energy-efficient construction long before the oil crises shook the world. When it faced a serious housing shortage after World War II, Sweden began an ambitious program to build a million affordable units. Interestingly, that program was launched at the same time the country experienced an energy crisis—a shortage of timber, its main fuel. As part of their early energy conservation crusade, write Lee Schipper and his coauthors in *Coming in from the Cold: Energy-Wise Housing in Sweden*, Swedish engineers developed innovative ovens, double-glazed windows, and ceiling and wall insulation that were to make them leaders in energy-saving technology decades later.

"Anyone who has endured a raw January day in Stockholm—where winters are much like those along the coast of Maine—can appreciate the Swedish obsession with quality housing," writes Schipper. "The snug, well-lit interior of Swedish homes provides a cheerful defense against the chill and gloom of winter." Yet, average energy consumption in the Swedish home is barely two-thirds that of other European countries and the United States. A survey of average oil consumption for space heating in single-family dwellings showed that, by the early 1980s, Sweden, Canada, Denmark, and Germany—countries that are farther north than the United States—all used less oil per square foot of floor area than the United States and France, according to Schipper.

Another way in which U.S. buildings fall behind is in the construction process. "We haven't progressed since the colonial days in terms of building processes," says Kando. "The Swedes can produce a better-quality house in half the time American builders can." Almost 90 percent of Swedish homes are assembled from factory-built elements, and maximum efficiency is the byword for production and transportation as well.

In a way, the U.S. building industry faces much the same problem as U.S. car manufacturers do. The Swedes have shown us that more fuel-efficient houses need not sacrifice comfort or style, just as Japan has shown us that fuel-efficient cars can be comfortable, reliable, and sleek. But houses last far longer, represent far larger consumer investments, come in far more styles, and are harder to import, so making the housing stock more energy efficient is a far more difficult challenge than improving the auto fleet's gas mileage.

Very recently, reports Howard S. Geller of the American Council for an Energy-Efficient Economy, energy efficiency in U.S. buildings has begun to rise. Although residential and commercial buildings used about 6 percent more energy in the first half of 1988 than they did in the first six months of 1987, the number of buildings grew during that time and temperature extremes and falling energy prices also led to increased energy use. According to Geller, increases would have been considerably higher if we hadn't made some efficiency gains.

PARADISE LOGGED

Another major source of carbon dioxide, the most important greenhouse gas, is deforestation. As more forests are felled for timber, wood burning or decay releases more carbon dioxide. Destroying vast forested lands also short-circuits Earth's natural water cycle. The forests of the humid tropics, in particular, pump prodigious amounts of water back into the air, produce clouds, and reflect back some of the sunlight that would otherwise dry that land. More important, the evaporation of water by trees removes heat from the entire region. In the aggregate, deforestation limits the planet's ability to recycle carbon dioxide. By reducing the global rate of photosynthesis, deforestation increases the fraction of yearly carbon dioxide emissions that stay in the air and releases the carbon that has been locked up in trees for decades or even centuries.

By the early 1960s, when the U.N. Food and Agriculture Organization completed a global survey, between one-third and one-half the world's original forests had been cleared. During the last few decades, the pace has quickened, and of the world's tropical forests alone, about one acre is now disappearing *every second*. Deforestation contributes an estimated 10 to 30 percent (most experts say 20) to the buildup of carbon dioxide. In some cases, it has certainly promoted desertification, and some scientists suggest that the general decline of ecosystems in parts of drought-stricken Africa, for example, may be the legacy of removing the trees that once grew there.

THE GLOBAL PERSPECTIVE

Today, developing countries account for only about one-fourth of the greenhouse problem, and historically their contribution was even less (about 10 percent). Yet, there has been a big increase in the use of fossil fuels—and CFCs—among many developing nations. The U.S. share, approximately 32 percent of global primary energy consumption in 1970, declined to about 24 percent in 1987. The Soviet Union's total rose from 14 percent in 1970 to 19 percent in 1987, while China's increased from 5 to 10 percent and the rest of South Asia doubled its share of the world total.

This shift is more complex than it seems because different countries use different mixes of fossil fuels and different fossil fuels have different carbon dioxide emissions rates. The United States, for example, uses more oil and gas (49 percent and 25 percent) and less coal (25 percent) than the world average. For the world as a whole, oil represents 43 percent of all fossil fuels used and coal, 35 percent. Worldwide, natural gas use accounts for about 22 percent of total energy consumption. Given these differences, much of the planning for fuel cutbacks and fuel switching will be a national affair.

In this country, one unfortunate consequence of the oil price shocks of the early 1970s was that we turned more and more to coal—which, to supply the same amount of heat, emits 72 to 95 percent more carbon dioxide than natural gas does. (Oil emits 38 to 43 percent more.) Electric utilities' use of coal rose by 75 percent between 1973 and 1987.

In the developing world, too, according to José Goldemberg and the coauthors of *Energy for a Sustainable World*, the emphasis has traditionally been on expanding supply and meeting the burgeoning demand

ENERGY EFFICIENCY: WHAT COUNTS?

Countries vary widely in the amount of energy they burn to earn each dollar of gross national product (GNP). Efficiency does not correlate exactly with size or wealth. Among industrialized nations, Switzerland is the most efficient. On average, the United States burns more than twice the amount of fuel per person as Switzerland does to achieve the same standard of living.

Country	Energy Index	GNP per Capita	Population (in millions)
Switzerland	100	$17,840	6.4
Japan	116	12,850	123.3
Argentina	116	2,350	32.4
West Germany	124	12,080	60.4
United Kingdom	140	8,920	56.2
Sweden	160	13,170	8.3
Australia	180	11,910	16.5
Mexico	224	1,850	87.0
United States	**224**	**17,500**	**246.3**
India	316	270	813.4
Canada	320	14,100	26.5

for fuel-fired transportation—instead of on using that energy more efficiently. Energy aid programs for the developing world have likewise emphasized energy supply, these authors write, with "90 percent of the expenditures going into large systems for generating, transmitting, and distributing electricity." As developing countries continue their historic transition from traditional vegetation-derived fuels to fossil fuels, their per-capita use of fossil fuels will surely rise. And as these countries' populations continue to grow, their total carbon dioxide emissions will too.

TOWARD A CONSERVER SOCIETY

In 1990, the United States still uses more fossil fuel than any other country does. Here, carbon dioxide emissions have three primary sources: electric utilities, transportation, and residential buildings and industry together. These each account for about one-third of all such emissions. The United States also ranks as the biggest user of chlorofluorocarbons, as well as one of the largest contributors to the buildup of three other greenhouse gases: methane, nitrous oxide, and tropospheric ozone.

In 1988, retail energy sales accounted for about 10 percent of the U.S. gross national product, but only 5 percent in Japan and Western Europe. Each country's industrial mix plays a role in determining energy efficiency in relation to GNP, but broadly speaking, Japan and West Germany now produce approximately twice as much GNP per unit of energy as we do—a double testimonial to these nations' efficiency and to our inefficiency.

Since the energy crises of the 1970s, energy efficiency has improved in some sectors of the U.S. economy. Jets and motor vehicles have become more fuel efficient. So have buildings. And many industrial processes now require less energy to make the same amount of goods.

Sources of Carbon Dioxide
Percent of U.S. carbon dioxide emissions in 1987.

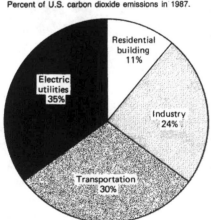

But these improvements have been offset by the growth in the number of cars and the number of miles that each is driven.

Following the energy crisis of the 1970s, new attitudes about conserving energy developed and new standards for energy-using devices temporarily took hold. For awhile, people turned down their thermostats in winter, turned off their air conditioners in summer, and drove (a bit) less. Energy-efficiency technology spread to residential and industrial heating systems. But the energy-conserving ethic eroded during the 1980s with the return of cheap oil and the dismantling of many government programs to encourage energy conservation and efficiency. The bad news is that we aren't as energy efficient as we would be had these changes not occurred. The good news is that we responded quickly to a perceived crisis at least once before and we can do so again.

3

Come Hell or High Water

Man did not weave the web of life—he is
merely a strand in it. Whatever he does to the
web, he does to himself.

CHIEF SEATTLE, 1854

In the Old Testament book of Genesis, the prophet Joseph listened to an
account of one of the Egyptian Pharaoh's dreams and warned that seven
"fat years" would be followed by seven "lean years." He urged the
Pharaoh to store grain as a hedge against the adversity to come. With
his dream interpretation, Joseph became one of the first to write a scenario for variable climate—as well as a plan for dealing with it. This
principle, dubbed "The Genesis Strategy" by climatologist Steven
Schneider, can help societies build a kind of safety margin into food
reserves.

Since Joseph's time, planning for climatic variability has become a
major concern not just of farmers and agronomists, but also of water-
resource managers, economists, and others. But the attempt to forecast
climate changes stemming from the buildup of carbon dioxide and other
greenhouse gases—changes that are certain to have far more dramatic
repercussions than any short-term weather event has had—is a new
business. Surely, a legion of experts besides those already on alert,
from oceanographers and geographers to political scientists and
sociologists, will have to be pressed into service to prepare us for such
pervasive changes.

Current theory suggests that if the warming effect of the buildup of
carbon dioxide and other greenhouse gases reaches the equivalent of
twice the preindustrial level (the fabled "CO_2 doubling"), the atmosphere will be "committed" to warm by 3 to 9 degrees Fahrenheit. At
the rate we are now going, that symbolic set point will probably be
reached before the middle of the twenty-first century, perhaps before
2030. If fossil fuel combustion were reduced and other complementary
measures were taken, that day, of course, could be delayed.

No matter what we do now, we can't avoid some future climate

50

change. There will be more than seven "lean years" ahead. Even if we were to stop all greenhouse gas emissions tomorrow, we still face an accumulated buildup from the last century whose physical effects won't be fully known at least until the end of this century. We don't know whether there will ever be "fat years" like those we used to know, or how societies can cope if there aren't.

What we do know is that, as at least one observer puts it, there's nothing green about the greenhouse problem. In fact, Norway's prime minister Gro Harlem Brundtland suggests calling it the *heat trap* instead. After taking stock of what scientists say could happen, the terms we now use do seem rather like bureaucratic euphemisms.

By whatever name, a global warming *will* increase urban ozone levels and the frequency of smog alerts. Scientists aren't predicting that people could change the atmosphere enough to create a living hell. But many are saying that the impacts of any rapid global warming will be far reaching and long lasting. The rest of this chapter is a tour of some of the possible impacts of a rapid global warming. Not all will necessarily occur. Some may be inhibited or aggravated by others, and some may be less severe than the worst scenarios suggest. But the economic costs of even scaled-down damages to human societies and natural ecosystems—which we can do something about—are likely to be tremendous, and the rate of climate change will determine just how serious those damages are.

FARMS, FOOD, AND FAMINE

Throughout history, some civilizations have budded and flowered only to wither and even disappear in response to climate change. One favorite example of sleuthing climatologists is the case of the long-lost Mycenaean empire. In *Climates of Hunger: Mankind and the World's Changing Weather*, climatologist Reid Bryson and writer Thomas Murray describe Mycenae as a telling lesson for our own time.

The ancient city of Mycenae now lies in ruins on a sunny plain 60 miles southwest of Athens, but in 1200 B.C., Bryson and Murray write, it "was the hub of a great civilization. Its massive main gate, with two stone lions on guard, its main walls, half a mile long and up to 30 feet thick, testify to the power it held. Its excavated tombs have revealed a wealthy and sophisticated warrior civilization with a farflung trade that dominated the Aegean and much of the Mediterranean seas for centuries." Suddenly, though, the empire began to decline. In 1230 B.C. the palaces and granaries of Mycenae were apparently attacked and burned, as other nearby cities—now remembered only in the legends of Agamemnon and Achilles—decayed.

Putting a new spin on Mycenae's decline and fall, classical scholar Rhys Carpenter argues that drought, not an invasion, was the root cause and that a diaspora of drought-stricken inhabitants, not conquerors, burned the overlords' palaces. Although sedimentary records yield little evidence, Bryson and Murray argue that a massive rainfall shift could have occurred in Greece, just as storm tracks shifted in the Mediterranean in the recent past—in January 1955, for example—disturbing the expected rainfall in winter, a crucial time for agriculture.

The emptied granaries of Mycenae dramatize what climate change could do. With rapid climate change, of course, perverse unpredictability would be the rule for agriculture; and the stakes are higher now because so many more lives around the world would be at risk.

Drought or dryness can come as a result of lower than normal rainfall, higher than normal temperatures, or both. With a global warming, temperature would rise and evaporation increase; at the same time, rainfall patterns could shift, a change that, in turn, could alter ambient temperature. In 1988, higher than normal temperatures in the United States combined with lower than normal rainfall to lower yields severely, for a record 35-percent reduction in corn alone.

Is this indicative of what might happen on the U.S. farm? Research completed so far points to serious dislocation if climate changes rapidly. As temperatures rise, the productivity of rainfed farms in the grain belt could fall. The Environmental Protection Agency (EPA) estimates that corn, wheat, and soybean yields could drop dramatically (by as much as 80 percent by 2050 or so) in such Great Plains states as Kansas, Nebraska, Oklahoma, and Texas. Elsewhere, too, grain yields would fall as temperatures and moisture stress rise.

Hot, dry temperatures, as the disastrous drought of 1988 proved, can devastate crops. Corn is particularly sensitive to heat during fertilization, which usually comes at the peak of the summer. If hot, dry weather persists throughout the 12 days of tasseling, the whole crop can be lost. Wheat will not mature at high temperatures, and preliminary evidence suggests that rice crops may also fail if temperatures rise much above 100 degrees Fahrenheit.

Higher temperatures and reduced rain could greatly increase demand for water in coming decades because the primary way to compensate for higher temperatures would be to irrigate more intensively. Farmers would need to change irrigation schedules, and demand for groundwater would escalate, putting supplies at risk. For instance, eight Plains states draw heavily on the vast Ogalalla aquifer beneath them to irrigate crops. But, the great aquifer—sometimes called the sixth Great Lake because it contains as much water as Lake Huron or Lake Erie—is already being seriously overdrawn. According to *Wilderness* magazine, the Ogalalla is being drained so quickly that "for every gallon of water pumped out, only a teacup is restored by the natural processes of aquifer recharge." Irrigation may also prove too expensive for most U.S. farmers, a Natural Resources Defense Council (NRDC) report stresses, "particularly those who are deeply in debt from purchasing land and machinery. A center pivot sprinkler system to water a half section (320 acres) of farmland, for instance, costs approximately $90,000 on today's market. Energy and maintenance costs compound the considerable expense of the machinery."

Even where rainfall would not be diminished, its patterns are predicted to change, though no one knows how. In congressional testimony in May 1989, James Hansen of the Goddard Institute for Space Studies, a scientist specializing in climatology, described how "a greenhouse-warmed climate is likely to have deeper, more intense thunderstorms with greater rainfall." As Gordon MacDonald writes in *Climate Change and Acid Rain*, "In Asia, the seasonal pattern of the monsoon rains on which the region's agriculture is dependent would be altered or interrupted."

A hotter, drier climate might also increase salinization problems. Warmer temperatures combined with increased irrigation mean a higher evaporation rate and greater salinity in the soil.

Even if the worst consequences—including the droughts, soil chemistry changes, pestilence, and blight featured in doomsday scenarios—never materialize, the economic consequences of reduced crop yields and the greater demand for already stretched water supplies will be enormous.

Even small fluctuations can be economically important. As Reid Bryson's research shows, "an increase of 1.8 degrees Fahrenheit in the summer temperatures in the northern Plains can reduce the gross dollar income of the spring wheat farmers by $131 million, and a modest 20 percent shortfall of rain can cost another $137 million."

To put potential losses into perspective, consider the "net worth" of U.S. agriculture. The American farm economy, as the 1988 EPA report on climate change notes, represented 17.5 percent of the gross national product in 1985 and accounted for some $42.6 billion in exports. According to the Council for Agricultural Science and Technology, agriculture also employs 21 million people (if both farm workers and those who process, distribute, and sell food are counted), more than any other American industry.

Whether caused by a greenhouse warming or not, intensive drought would undoubtedly displace more and more farm families in the Midwest. Although some agriculture experts are optimistic that grain belts can simply "shift northward," critics point out that longer growing seasons farther north won't do the trick if the soils in these new agricultural areas are inferior. Many soils found in the northern humid regions of the Great Lakes are too acidic and not fertile enough to grow grain, which NRDC researchers say, "explains why large areas of cut-over forest lands in the northern Great Lakes region have never been put under cultivation."

Even if crops and farms could shift, would farmers shift with them? The Joad family in John Steinbeck's *The Grapes of Wrath* will symbolize forever the fate of those who were "dusted out" of the southern Plains states, but most families living in the Dust Bowl chose to remain. Agricultural historian R. Douglas Hurt notes that the vast majority chose to stay "because they had an overwhelming faith in the future. Three words—'if it rains'—dictated much of their daily life."

Droughts are also hard on livestock, which—like human beings—suffer heat stress. Preliminary studies suggest that high temperatures reduce fertility in cows, pigs, and other farm animals. Pastures and rangelands are also damaged by the heat, and crop losses diminish stores of animal feed.

More bad news: heat waves and droughts also help pests and diseases propagate, especially in crops and livestock weakened by heat and moisture stress. According to the NRDC report, "In 1988, dry conditions fostered proliferation of *Aspergillus flavus*, a fungus that produces an extremely potent carcinogen called aflatoxin that may have tainted as much as 6 percent of the year's corn crop. Furthermore, soybeans last year were ravaged by the spider mite, which caused an estimated 15 to 20 percent of the crop loss in some areas."

Heat brings greater concentrations of tropospheric ozone, a plant toxin that is causing losses of 5 to 10 percent in crop productivity nationally. Elevated ozone levels, especially in rural areas, amplify the negative impacts of acid rain on crops.

Another potential need for agriculture is finding crops that can weather the shift should climate change rapidly. Over time, farmers select crops to get the highest yields under average climatic conditions. If the climate changes, cultivating techniques and even crops may have to be changed. How successful a new regime will be and how long it will require to take hold are matters of speculation. "In principle, we can use new crop varieties that are better adapted to a changed climate," write William Kellogg and Robert Schware in *Climate Change and Society*. "But plant geneticists are concerned about the rapidly decreasing stocks of vigorous wild strains and disease-resistant food crops; these are being pushed out as their natural ecosystems are damaged and eliminated."

Drought could weaken the U.S. role as the world's breadbasket. If, as the National Academy of Sciences considers likely, "agroclimatic zones" shift toward the poles and other regions of the world find their growing seasons lengthened, the whole U.S. farm economy could eventually be threatened. "The United States could become a grain importer and the U.S.S.R. could become a grain exporter," Walter Orr Roberts of the National Center for Atmospheric Research reports in *Fortune* magazine. And even to the new breadbasket regions, such a shift would not be the godsend it might initially seem, because it would take years for agriculture to find new crop mixes and adjust in other ways to change.

If climate did change quickly, what would happen to supplies sent to much of the developing world and some industrialized countries, including Japan? In times of normal weather, the United States typically harvests more than 300 million tons of grain, consuming roughly 200 million of this total and exporting the surplus. According to the Worldwatch Institute, "if this surplus were to be substantially diminished, it could have severe consequences for the rest of the world dependent on the U.S. grain reserves." If another drought as disastrous as the one in 1988 or droughts in two or more consecutive years hit, "U.S. grain exports would slow to a trickle." In a scenario involving a "frantic scramble" for supplies, world grain prices would soar to record levels. "Asian and African countries, in particular, would find it impossible to feed their people without [U.S. cereals]," says the report, "as would such major cities as Leningrad, Cairo, Lagos, Caracas, and Tokyo, (which depend largely on grain from the United States and Canada.)"

Changes in agricultural yield can produce a huge number of social and economic ripple effects throughout the world, many of them difficult to predict. In 1972, drought forced the Soviet Union to purchase 18 million tons of grain from the United States and another 12 million in 1975, events that strongly influenced food prices and the U.S. balance of payments. Today U.S. grain reserves are at the lowest point in decades—with what consequences? Who knows for certain?

Of course, worst of all, the human toll can be staggering. In the late 1960s, the world began to take notice of the impact of widespread drought afflicting the semiarid region that spans six African nations. By 1973, write Bryson and Murray, twenty million people were in the grip of a drought that "destroyed their pasture lands and their grains, dried up their wells and rivers, killed over a third of their cattle and more than 100,000 people." Only food shipments from other parts of the world saved the millions when their summer monsoon rains failed. By most reckonings, that drought lasted at least another decade, and, as analysts have concluded with hindsight, overgrazing and other poor environmental planning exacerbated the situation.

In the last decade, many parts of the world have been stricken by droughts that took lives and a heavy toll on the international economy. Droughts in 1972 and 1973, triggered by El Niño, the periodic warming of sea-surface temperatures in the eastern equatorial Pacific, devastated a wide range of farming regions—in the Soviet Union, China, India, Central America, Australia, Indonesia, Argentina, Ethiopia, and elsewhere. The 1982 and 1983 El Niño–related weather disasters struck even harder.

James Hansen's global circulation models show that droughts due to global warming would have the largest impacts in certain parts of the subtropics and middle latitudes, particularly in the United States, Mexico, southern Canada, southern Europe and the Mediterranean, the Soviet Union and China, the African Sahel, Australia, the southern quarter of Africa, and parts of Brazil and Argentina.

Droughts have jarred the world into recognizing just how precarious world food security can be. Twenty years ago, writes social scientist Michael Glantz, world leaders were confident that "society had developed the technologies needed to buffer agricultural production from the vagaries of climate and, as a result, the world was about to enter an era of cereal grain sufficiency, if not overabundance." Now decision makers and the public have been jolted, he says, "from their complacency regarding their immunity from the impacts of the vagaries of weather."

A drier future unfortunately appears likely. The National Research Council's Board on Atmospheric Sciences and Climate indicated in 1987 that recurrence of the recent summer drying and warming experienced in the mid-latitudes would be even more "likely in the long term." As James Hansen states in a letter to the *New York Times*, "[NASA's] climate model, tested by simulations of climate on other planets and past climates on earth, indicates that the greenhouse effect is now becoming large enough to compete with natural climate variability." "The model," he goes on, "gives us a hot summer in the United States in the 1990s about 60 percent of the time, compared with 33 percent of the time for 1950–79. Droughts in the United States increase similarly."

Not every region will get drier if temperatures climb, and not all the effects of climate change on agriculture would be adverse. Globally, rainfall is expected to increase in a warmer world, and if some of this rain falls in traditionally dry regions, existing deserts might bloom. Some models predict more consistent monsoons for India and Sahelian Africa, for example, and land in cold, high-latitude regions—including Canada and the Soviet Union—could become more suitable for production of some crops. Similarly, as carbon dioxide levels rise, plants could grow bigger. In the last instance, though, the benefits may be offset by a peculiar chain of events. Rapid plant growth saps soils of nitrogen and other nutrients, making the protein content of leafy plants decline, so insects would have to feed more voraciously to get their fill of nitrogen. With hungrier pests and faster-spreading diseases under warmer conditions, farmers would be tempted to spray more pesticides, some of which have high economic and health costs.

Recent events foreshadow in a small way the possible impacts of a greenhouse warming. Dry, wilted crops in the Midwest, forest fires in the Southwest, starving cattle slaughtered in the Great Plains, and ravaged harvests of corn and soybeans are all legacies of the drought of 1988 and its toll on American agriculture. Although this drought has not been linked to a warming trend already in effect, it gives us some snapshots of what could be in store for us.

SEA LEVEL—THE GREAT EQUALIZER

Many unknowns surround climate change, but the inevitability of sea-level rise isn't one of them. Scientists know for certain that the average global sea level will rise with increasing temperatures even though they don't know exactly how great the rise will be. During the past few thousand years, the seas have been relatively high and stable, after hundreds of millennia of 300-foot variations. During various ice ages, temperatures were 9 or more degrees Fahrenheit lower and sea levels were more than 300 feet lower than they are today.

In the comparatively recent past, sea level has risen by only 4 to 8 inches per century. But ocean rises associated with a global warming may not be as comfortably gradual and modest. Greenhouse gases that have accumulated in the atmosphere could raise the sea level globally anywhere from 1 to 4 feet during the next hundred years, according to Roger Revelle at the University of California, San Diego.

The mechanism underlying sea level rise is simple. As ocean waters heat, they will expand. Water locked up in small landed glaciers will melt. If the warming is at the high end of the range predicted, during the next few centuries it could even melt the polar ice cap. The Greenland and West Antarctic ice sheets each contain enough water to raise sea level about 22 feet, while East Antarctica has enough ice to raise sea level over 190 feet, according to the U.S. Polar Research Board.

According to James Titus of the Environmental Protection Agency, a 1-foot rise in sea level—the rise anticipated by 2075—would allow seawater to penetrate some 35 yards inland along a coastline of average steepness. But a 3-foot rise would flood low-lying coastal areas to about 1,000 yards inland. This projected rise in sea level, as this process accelerates, would exacerbate coastal flooding in such places as Taipei, Venice, and the Nile Delta. In the United States, most of the Atlantic and Gulf coasts would suffer. Especially hard hit would be the Carolinas, where the slope of the shoreline is so gradual that a 1-foot rise in sea level could push the shore back miles.

Louisiana's prospects may be the most unsettling of all. According to EPA's Titus, the entire port of New Orleans might have to be moved. Southern Louisiana has already lost large dryland areas to erosion, inundation, and other causes; roughly a million acres of wetlands have been lost since 1900. By 1995 the annual wetland losses alone are projected to reach 120 square miles. To a region that supplies about 30 percent of the nation's fish catch and 40 percent of its fur and that also provides a winter home for about two thirds of the migratory birds in the Mississippi flyway, accelerated damage to drylands will come as a crushing blow.

Miami, Florida, is another vulnerable city and the subject of a recent study by the Urban Institute for the EPA on what a 3-foot sea-level rise could do. Most of its greater metropolitan area lies at or only slightly above sea level, with the groundwater only feet below street level. What's more, the city rests on highly porous limestone rock and soil already prone to flooding.

Miami's first settlers built on high ground surrounded by the Everglades, the lush mangrove swamp that once covered the entire southern tip of Florida and is now protected as parkland to the west. As the city grew, the swamp was drained and reclaimed. But Miami still draws water for its 3 million residents from the Biscayne aquifer, one of the world's most permeable, which lies just a few feet below its city streets. Were it not for what engineers call a "hydrologic masterwork" of canals and levees —an extensive, intricate drainage system that provides fresh water, irrigates farm land, limits the "saltwater" front pushing into the aquifer and rivers, and holds back swamp and sea—the city would be under water.

How, under such circumstances, could Miami possibly cope with the 3- to 5-foot rise in ocean level anticipated to occur by the year 2050? The answer in The Netherlands, as well as in New Orleans, has been to build very expensive dikes. But the Urban Institute argues that even a system of dikes and seawalls would not be enough to save Miami.

According to the Urban Institute's report, "One would have to construct a water-impermeable barrier along the length of Broward and Dade Counties to a depth of 100 to 150 feet. Without such a wall, the pressure of the sea water would cause the sea to rush into the aquifer below. . . ." If sea water were to wash in, roads would buckle, bridges sink, and land revert to swamp.

Instead of building dikes, the Urban Institute suggests the other logical alternative—raising the land in low areas. This would mean raising all the city streets in Miami by 3.3 feet, reconstructing causeways, and raising some canals and levees. The result, says the Urban Institute's Ted Miller, would be to transform the flat look of Miami completely. Houses, yards, and garages would sit several feet lower than the streets. "People will stand in their front yards and look up at the street," says Miller. "It will look something like lowland areas in the Dutch countryside."

Coastal geologist Stephen Leatherman questions whether dikes might not suffice for Miami as they have in Holland for centuries: "As the Dutch say, 'God made the Earth but the Dutch built Holland.' " So far, though, the Urban Institute study is the only exhaustive report of its kind. And no one questions that the cost of saving Miami from the onrushing tide will be exorbitant. The Urban Institute estimates roughly $237 million will be needed to raise the streets alone, with an estimated total of about $600 million in 1987 dollars for raising levees and canals, altering drainage, and modifying the airport. The costs of raising bridges, pumping water, and changing the water supply system to cope with saltwater intrusion would be additional. Also left out of these preliminary estimates are the substantial private costs of raising houses and yards that lie too low, adjusting sewage pipes, and building trenches—and, of course, building all those steps up and down between houses and streets.

Miami would be just one city around the world devastated by the effects of sea level rise. But while its peculiar hydrology is unique, its problems with flooding, intruding seawater, and storms won't be. New York, Los Angeles, Shanghai, Buenos Aires, Venice, Tokyo, Bangkok, Taipei, and the other seaside and estuarine megalopolises, where well over half of humanity is clustered, all face similar problems. The costs of additional damages to such densely populated areas seem almost incalculable. Maintaining the threatened shorelines on the U.S. East Coast alone will cost an estimated $10 billion dollars if sea level rises 3 feet due to climate change, according to the EPA.

Another casualty of sea-level rise will be the world's barrier islands that protect the mainland and enclosed lagoons from wave attack by the ocean. Most of their shores have retreated during the last century, and sea level has risen by a worldwide average rate of 6 inches. Seventy percent of the world's beaches have eroded over the last century, but coastal experts estimate that sandy beaches will erode two to five times faster if sea level rises further. U.S. beaches have been most prone to erosion—most eroding in a range of 50 to 100 feet in the last 100 years along the East Coast.

"The United States has the longest and best-developed chain of barrier islands in the world," writes University of Maryland coastal expert Stephen Leatherman in *Barrier Island Migration*. Favorites of tourists and vacationers, they dot the eastern seaboard from Maine to Florida and the shores of the Gulf of Mexico. Unfortunately, though, the U.S. barrier islands are eroding particularly fast. "Along the Atlantic coast, beach erosion has historically amounted to about two to three feet per year, with the Virginia barrier islands exhibiting the highest rates of erosion (tens of feet per year)," Leatherman testified before the U.S. Senate in 1988.

One of the main shaping forces for these weather-beaten islands in recent times has been beach development. Some of the islands are so impermanent and prone to erosion that it's a risky proposition to build a house, let alone condominiums and resorts. But others, such as Maryland's Fenwick Island—now known as Ocean City, the host to 8 million visitors each year—is overbuilt with multistoried condominiums and hotels. To keep up with shifting sands, officials have spent millions to pump sand onto the beaches and to rebuild natural dunes; they may end up spending billions more to keep buildings from sinking.

Beach erosion as a greenhouse problem has yet to capture public attention the way the 1988 drought and heat wave did."Hopefully," writes Leatherman, "a coastal disaster along an urbanized beach will not be necessary to promote public awareness of the sea-level rise phenomenon and its attendant impacts."

Beach erosion destroys shorelines at tremendous cost. But losing the fragile coastal wetlands that fringe the world's shores, with their rich stock of birds and aquatic creatures, may be even more costly. Comprising much of the land less than 5 feet above sea level, coastal wetlands form the breeding ground for many of the world's fish and shellfish. The marshes, swamps, and mangrove forests along most of the U.S. coastline, particularly along the Atlantic and Gulf coasts, for example, are vital for shrimp, crab, oysters, and many fish, and they are natural filters for water pollution. They are also nesting ground for hundreds of species of birds and waterfowl, as well as home to many amphibians and some mammals. "All in all," writes Norman Myers in *The Greenpeace Examiner*, "the demise of coastal wetland could prove to be the greatest wildlife-related impact of the greenhouse effect in the United States."

Of course, only a small fraction of the world's vulnerable coasts are American. Worldwide, some 70 percent of all beaches are already eroding under pressures from off-shore dredging, residential and industrial development, and land filling. Nations such as Egypt and Bangladesh would be particularly hard hit because their populations have clustered on river deltas, where flat, fertile, low-lying land has built up from sediment washed down from major rivers. Some 15 to 21 percent of Egypt's 51 million inhabitants live on lands that would be swamped by a 1- to 3-foot rise in sea level, according to economist James Broadus of Woods Hole Oceanographic Institute. The toll in low-lying Bangladesh, where severe flooding is common, would be even worse. If a 3-foot rise occurred, the already overcrowded nation would lose 12 to 28 percent of its total land area and would be visited by more floods and tropical storms like those that killed 300,000 people in the early 1970s and tens of thousands of people in 1988. The Ganges-Brahmaputra-Meghna delta (the portion of Bangladesh that juts into the Bay of Bengal) may witness the worst devastation. "It's massively populated, achingly poor, and something like a sixth of the country is going to go away," Robert Buddemeier of Lawrence Livermore National Laboratories warns in *Discover* magazine.

Throughout the southern oceans, too, damages would be hard to prevent. Areas in Australia, for example, might be able to hold back the sea with dikes and dams, writes Eric Bird in *Coastland Changes: A Global Review*, but many New Guineans live in small cottages near the water's edge on barrier islands so vulnerable to erosion that it would make little sense to build a dike. In the Philippines some have given up fighting the sea. Here, "many people have literally 'taken to the water,' living in small boats and maintaining fishing nets in their own plots of bay instead of land," according to a 1986 report by the EPA and the United Nations Environment Programme.

Ocean resorts in vulnerable areas around the globe will fare better, of course, than the more sensitive coasts. The financial investments in such at-risk developments are easier to quantify and defend than are, say, potential losses of marine habitat or future near-shore fish catches. According to Leatherman, who studied how sea-level rise would register in South America, such resorts as Copacabana Beach, Brazil; Punta del Este, Uruguay; and Mar del Plata, Argentina, probably have enough clout to convince their respective governments to allocate funds to protect them, but "coastal wetlands will receive benign neglect."

For many coastal areas and wetlands, sea-level rise will be so disruptive as to raise the question of whether it might not be better simply to move inland with the oncoming tide. In one study, scientists projected that 40–75 percent of U.S. wetlands could be lost by the end of the twenty-first century. Those losses would be reduced by as much as half, however, if people inhabiting developed areas simply vacated them as the sea level rose—allowing the wetlands to be naturally replenished—rather than constructing bulkheads to protect their property. According to Duke University coastal geologist Orrin Pilkey, "Retreat is the ultimate solution."

Yet another far-reaching effect of the greenhouse warming on the world's oceans could be an increase in sea-surface temperatures. Satellite reports in 1989 suggest that surface temperatures may already be rising at a rate of 0.4 degree Fahrenheit per decade, though the El Niño current may be causing the warming. If the top layers of the oceans do get warmer, we can expect more tropical cyclones, hurricanes, and other storms. As wave action and cyclones gather force, a sea-level rise could also invite further erosion of beaches and the destruction of waterfront property.

Such an increase in sea surface temperatures could kill tropical coral reefs—the world's most diverse ecosystems and possibly the oldest as well—and the animals and plants that live on them; a Neptune's locker of marine organisms could be reduced or extinguished.

For reasons still debated, the 1982–1983 El Niño—the periodic warming of sea surface temperatures in the eastern equatorial Pacific—rocked weather patterns, leaving thousands dead and hundreds of thousands on at least four continents with vivid, nightmarish memories of what unpredictable weather events can mean. Writing in *National Geographic* in 1984, Thomas Y. Canby toted up some of the ravages: fatal floods and coastal landslides in Ecuador, starvation and mass migration in drought-stricken southeastern Africa, smothering dust storms covering 77,000 square miles in Australia, cyclone blasts in

Polynesia, record rains and collapse of fisheries in California, and more. Estimated at some $8.7 billion, these damages are associated with a regional weather change rather than global warming, but they remind us at once of how vulnerable coastal life is, how ineffective conventional disasterproofing can be, and just how large and imperfectly understood are the forces of nature that we are tampering with.

WATER WARS

Throughout history, water has been fought over for its importance as an economic resource and political tool, as William Kellogg and Robert Schware point out. These experts believe that lack of precipitation may pose greater problems than the temperature change itself, jeopardizing food production and the management of forests and rangelands.

The Soviet Union's recently abandoned plan to divert water from the Ob River south to the vast desert in the Kazakhstan and Uzbekistan regions was an incredibly elaborate scheme for bringing new water supplies to agriculture. "Scheduled for completion in the year 2000," writes Walter Orr Roberts, "the project [was to] reduce the vulnerability to climate fluctuations of Soviet grain production." The monumental project would have carried the river's 88 trillion gallons of water per year through 1,500 miles of tunnels, canals, and pumping stations. Whether the plan is revived, the scale of the project demonstrates the value placed on water in drylands and hints at the herculean efforts that might be needed to keep agriculture viable as more soils lose their moisture in a greenhouse-altered climate.

In the Middle East, a water fight is in the making between Egypt and Sudan, both of which draw water from the Nile River and both of which have rapidly growing populations. Sudan wants to divert a larger share of the river's water; Egypt wants to hang on to what it has. In 1988, Egypt's President Mubarak plainly indicated that if the Sudanese attempted to reduce the flow from the Nile's headwaters into Egypt, the move would be construed as an act of war.

Equally intense competition for scarce water resources is likely to emerge in the next few years between Israel and Jordan, both of which depend heavily on the waters of the Jordan River.

In the United States' arid west, there have been long-standing water disputes over the Colorado River, but the arguments have quickened in recent years as population growth and development dramatically increase the demand for water. Because the rate of water evaporation will rise with rising temperatures, demand for water in the west—already on the rise for other reasons—can only intensify. Water will be scarce for irrigation, which depends on having enough runoff after evaporation. "With a warmer and drier climate . . . and with a reduced supply from the Colorado River as Arizona utilizes its water rights under the Colorado River Compact and as Indian water rights come to be more utilized," water supplies could come up short, Roger Revelle told a congressional committee on water and power resources.

In California, where some 78,000 farmers raised close to $16 billion in agricultural produce in 1987, nearly half the water comes from the Sacramento and San Joaquin river basins, fed by snow runoff from the Sierra Nevada and other mountains. Normally, the state keeps reservoirs low in the spring, to guard against floods, and high in summer, to supply the thirsty Central Valley, with its massive farming operations, and densely populated southern California. Climate change, says Peter Gleick of the Pacific Institute for Studies in Development, Environment and Security, would disrupt California's fine-tuned water supply management technique. If there were more rain and less snow falling in winter, the state might witness burst dams and devastating floods, Gleick warns, while summer supply levels could dwindle because of drought. Gleick told a congressional panel that this situation would give California "the worst of all possible worlds . . . more severe winter floods . . . [but] sooner summer drying and higher agricultural water demands." And Robert Buddemeier points out that California's population could double by the time these predictions materialize, intensifying urban demand for water and rural demands for irrigation.

Wintertime floods and summertime droughts won't be California's only problem. Saltwater pollution will also increase. As ocean levels rise—and they have already risen enough to infiltrate freshwater aquifers beneath the Sacramento River delta—seawater will push brackish waters farther inland, endangering existing water supplies. "Under present conditions, water quality deteriorates somewhat during low-flow periods of late summer and early fall," Buddemeier has reported to a congressional committee. "During drought years, salt intrusion and water quality can become a serious problem."

Changes in freshwater runoff from the Sierra, Peter Gleick has testified, would not merely alter the natural flushing of salt water from the delta. "Perhaps more important, rising sea level will push salt water farther and farther into the delta, threatening both water quality at the delta pumping plants and the delta levees themselves," he says. According to an EPA study, vast regions of the delta could revert back into the inland marshlands they once were unless very costly measures were taken. How costly? "The expense of maintaining all the levees may well exceed the value of the agricultural lands they protect," Gleick speculates.

If California's water supply were threatened, it would hurt the quality and availability of much of the fresh produce U.S. food shoppers take for granted—from California oranges and grapes to salad greens and nuts—and force us to import more foodstuffs from Europe and elsewhere.

In the Midwest heartland, a predicted decrease in rainfall and rise in temperatures could cause water levels to drop drastically in the Great Lakes area, a huge region whose shipping and manufacturing industries account for one-sixth of the gross national product. If levels dropped enough, deep-draft freighters would be blocked from the lock-and-dam systems, which could drive up costs for shipping vital cargo to such cities as Chicago, Detroit, Toronto, Buffalo, and Milwaukee.

The Mississippi River's water levels hit historic lows in 1988, stalling barge traffic from New Orleans to the Great Lakes. If droughts persist, such troubles—which were not predicted to attend midcontinental drying—could become more prevalent after the turn of the century. In 1988 in the Memphis area, for example, about 1,000 barges were stalled along the river; meanwhile, downstream of those barges, an Arkansas power plant had to be shut down. According to the *New York Times*, "its water intake pipe, normally under 20 feet of water, was clearly visible." Had the drought continued, eight water-cooled nuclear power plants might have been forced to shut down for an indefinite period, according to the Nuclear Regulatory Commission.

On the East Coast, saltwater intrusion could be even more troublesome in places like the Delaware River basin. A substantial sea-level rise there, fears Bob Alpern of New York City's Citizen Union (a group devoted to water planning), "could produce so much erosion and so much seepage of seawater through the soil that some portion of the Long Island aquifer could be hurt." As a result, saltwater intrusion could turn many domestic wells brackish, he says. Philadelphia, which draws its water from a tidal river, and New York would, like California's Central Valley, be particularly vulnerable.

But the changes go far beyond impacts on urban societies. As rising sea level pushes seawater upstream into rivers, it could upset delicately balanced estuaries. If that happens, Alpern points out, precious freshwater supplies will have to be released to counteract the imbalance. These reservoirs will already be overtaxed if the global warming reduces the rainfall that feeds them.

Water—the lack of it, the way it's distributed, the struggle to control it—has historically shaped the United States, especially the western states. Climate change would harden battle lines because so much money has been invested in the current water system and any change will be expensive for at least some water users. Too much water in some regions—caused by rising sea levels and more frequent storms—could cause disastrous flooding in coastal regions and river deltas. Too little water in other regions—the result of shifting rainfall and snowfall patterns—could reduce agricultural yields and increase political conflicts. On the other hand, some long-dry areas might see an increase in the rains whose lack had prompted decades of prayers. That some might win while others lose is small consolation. What would be most troubling would be the persistent uncertainty about the regional distribution of effects. Who could know, for example, whether downtown Baltimore might not be flooded in a torrential storm or the New Jersey Turnpike rendered impassable, or whether the West might not be prompted to buy expensive water from Canada?

SPECIES LOST IN THE SHUFFLE

Of all Earth's resources likely to be affected by climate change, natural ecosystems, including vast stretches of both forests and wilderness, stand to suffer the most. EPA's 1988 report on climate change put it bluntly: "If current trends continue, it is likely that climate may change too quickly for many natural systems to adapt." Less matter-of-factly, Harvard biologist E.O. Wilson calls possible species extinctions "the folly our descendants are least likely to forgive us."

How great might those losses be? That depends on one's geographical point of reference. Animals and plants living at middle and higher latitudes, most scientists believe, may be forced to move hundreds of miles toward the poles (or thousands of feet into the mountains) to find suitable habitats if temperatures rise only a few degrees. Large mammals and migratory birds may find the move easier than other species, provided they aren't hemmed in by mountains, lakes, or urban sprawl's shopping centers standing in their paths. Plants, on the other hand, can travel only as far as their seeds disperse—anywhere from a few feet to several miles with each generation.

How climate change would transform the tropics is less certain. Scientists expect smaller shifts in temperature and rainfall around Earth's girdle than in the temperate regions, but warming could spell disaster for many tropical species in other ways, report Walter V. Reid and Kenton R. Miller of the World Resources Institute. Even if the total amount of rainfall remains fairly constant, seasonal changes—wetter rainy seasons and drier dry seasons—could mean more hurricanes, forest fires, and other natural disasters that would alter species' habitats. And, as noted, rising sea level would cut into coastal wetland habitats and could drown coral reefs.

Whereas much of the world's agriculture depends on just a handful of plant species that can be rotated or even switched, natural ecosystems are too complex to be replenished or replaced so easily. Forests and wilderness areas, for example, support a huge number of diverse species living in a natural but delicate balance.

None of the world's ecosystems remains as pristine as Eden before Adam. In 1981, Kellogg and Schware singled out the inner reaches of the tropical forests and the unpopulated tundra as the few untouched areas on Earth. Since that writing, an estimated 165 to 250 million acres of tropical forests have been logged or converted to other land use, while oil drilling has proceeded apace in some Arctic areas. But even these comparatively pristine zones may be transformed by human-made climatic shifts. Tundra lands may come in for extreme temperature rises because of their high latitude, and tropical forests may suffer from changes in the timing and amount of rainfall and the frequency of storms and fires.

With greenhouse gas-induced warming, climatologists believe that some ecosystems would be able simply to "migrate," though it may take several centuries for the shift to occur. During the cooler, drier climate prevailing 3,000 to 4,000 years ago in north central Canada, for example, the Arctic tundra invaded spruce forests some 180 to 240 miles to the south in a relatively short time. Evidence abounds of animal and plant migrations in the warm interglacial periods, between the ice ages of the Pleistocene epoch, when North America was apparently 4 to 6 degrees Fahrenheit warmer than now. According to Robert Peters of the World Wildlife Fund, "Osage oranges and pawpaws grew near Toronto, several hundred kilometers north of their present distribution; manatees swam in New Jersey; tapirs and peccaries foraged in Pennsylvania; and Cape Cod had a forest like that of present-day North Carolina."

When the last ice age ended some 10,000 years ago, Reid and Miller add, the area now known as Pennsylvania was covered by pine and hemlock forests and inhabited by such small mammals as the smokey shrew, ground squirrel, and collared lemming. As Earth warmed and the glaciers retreated, the pines migrated northwest and the hemlocks headed northeast into New England. The warmer temperatures also drove the collared lemmings north, leaving William Penn's namesake lands inhabited by the shrews and squirrels. Elsewhere in the country, collared lemmings shared their ice age habitats with armadillos, but today the onetime neighbors live almost a thousand miles apart.

Seven thousand years ago, the prairie expanded in the American Midwest during a global warming episode, and many species followed. Some vegetation belts survived a move of hundreds of miles north. But other plant and animal species perished en route, because they either migrated too slowly or encountered such geographical barriers as oceans, mountains, or the wrong soil type. For example, a number of diverse plants that still survive in North America met with extinction in Europe during the Pleistocene ice ages. "Presumably," say Peters and Joan Darling, "the east-west orientation of such barriers as the Pyrenees, Alps, and Mediterranean" blocked their southern migration.

Today, cities would also be in the way. Peters notes that "Few animals or plants would be able to cross Los Angeles on their way to the promised land."

One significant threat to species would be the speed of climate change. As Peters puts it, real trouble begins "if present habitat becomes unsuitable faster than new habitat can be colonized." Plants now living near their thermal tolerance may have to shift northward at unprecedented rates to survive. "Although some species, such as plants propagated by spores or dust seeds may be able to match these rates," says Peters, "many species could not disperse fast enough to compensate for the expected climate change without human assistance, particularly given the presence of dispersal barriers."

Animals, unlike plants, are mobile, but their success in migrating would depend partly on whether the plants they eat survive. Then, too, some animals just aren't programmed to migrate. Beetles—the largest and most ubiquitous order of insects in the animal kingdom—are hardy itinerants. Indeed, their fossil remains have been used to track climate changes occurring over the past 50,000 years because they habitually strike off for more agreeable climes whenever they don't like the weather. But at the other extreme are some species of unadventurous deer that don't move much farther than 2 kilometers per year. Nor would many tropical birds that live deep in the forest even cross a clearing in the trees.

Given both these biological and geographical constraints to species' migration, conservationists fear that many species will be at risk. According to Peters, "By the next century most other surviving terrestrial species (other than humans) may well be relegated to small patches of their original habitat, patches isolated by vast areas of human-dominated urban or agricultural lands."

Without heroic measures to conserve habitats, say wildlife experts Thomas Lovejoy of the Smithsonian Institution and Norman Myers, hundreds of thousands of plant and animal species could become extinct by the end of this century, with more to follow in the next. Tragically, these victims of rapid climate change could disappear before they are even identified or counted, much less evaluated for use in medicine, industry, or agriculture. Walter V. Reid comments: "Some 230 years after the Swedish botanist Carolus Linnaeus began classifying the variety of life on Earth, we still do not know how many species exist—even to within a factor of ten—and of the estimated 10 million or more species on Earth, only some 1.4 million have been given names and at least a brief description. Up to 25 percent of the world's species alive in the mid-1980s may be extinct by the year 2010 or so, and significant losses of genetic diversity of both wild and domestic species are expected over this same period."

By almost any reckoning, a rapid global climate change could greatly reduce Earth's biological diversity—the planet's natural endowment of species and genes that is already being depleted at accelerated rates. "If the greenhouse warming occurs, it will pose a new and major threat to species within reserves, species already stressed by the effects of habitat fragmentation," says Robert Peters. Climate change, he declares, brings new pressures and new competitors, and it could quickly turn stress into distress. If climate changes quickly, these species' ranges and population sizes would shrink further, and some small populations might disappear.

Reid and Miller agree with Peters, noting that both the size and speed of climate change would hit species in temperate regions hard. When push comes to shove, they add, conserving biological diversity will rank low on the list of emergency needs brought on by global warming: out of self-interest, people will protect agriculture, water resources, and coastal development first. Both the money and land needed to maintain preserves and establish new ones will be in short supply, they assert, despite the multiple benefits of conserving biological diversity.

Even where survival is not the issue, "quality of life" may be. As biologist David Ehrenfeld puts it, most of our "energetic environmental activities bring . . . plants and animal communities back to earlier stages . . . dominated by organisms in conflict with people—the weeds, the pests, and the vermin." At a minimum, it takes decades to coax the original occupants back to beleaguered ecosystems, if they can be coaxed back at all. "In the meantime," says Ehrenfeld, "we must live with bamboo, the imperta grass, the bramble thickets, or whatever."

FORESTS AND THE CARBON CYCLE

Like the forests that "moved" from Birnam Wood to Dunsinane in Shakespeare's *Macbeth*, forest species have migrated in the past in response to climate change. Margaret B. Davis of the University of Minnesota notes that Eastern Hemlock, for example, "survived the last glacial period in the eastern Appalachians or on the coastal plain or shelf. Starting 10,000 years ago, the species moved northward at an average rate of 20 to 25 kilometers [12 to 15 miles] per century." By 5,000 years ago, the species had colonized all the eastern peninsula of Michigan and the northern half of the lower peninsula. Future climate change would again prompt migration of forest ecosystems. The question is whether they have the ability to adapt rapidly enough to survive.

Unfortunately for some forests—the boreal forests of Canada and Siberia, for instance—survival is a long shot. "If average global temperatures go up 5 degrees Fahrenheit, the boreal forests and their habitats will disappear completely," says William Moomaw of World Resources Institute. With the maximum migration rates of trees in mind, scientists predict that many species of trees would have to migrate up to ten times faster than normal to adapt to the kinds of temperature increases predicted. Other scientists say that the prairie-forest border now south and west of Minneapolis would have to migrate north at a rate of between 60 and 95 miles per decade to keep up with projected climate change, or from 250 to 375 miles by the year 2030. Although many "prairie" species—mostly grasses introduced by settlers—could easily make the move, many tree species couldn't.

Predictions are that the warming will strike the middle and higher latitudes most severely. According to Moomaw, if the average warming is 5 degrees Fahrenheit, temperatures could increase 12 or 13 degrees Fahrenheit in polar regions. As a result, boreal forest species such as spruce and Douglas fir in the northwest and spruce, balsam fir, and hemlock in the northeast would be likely to disappear completely— along with all the wildlife that depends on them for food or habitat. In the deciduous forests and mixed deciduous forests of North America, the ranges of many common trees, such as sugar maples, birch, beech, and some pines and spruce, could shift to the north.

But trees run into the same problems that other migrating species do, and whether they could just pick up and move is questionable, forest experts say. Trees that take longer to reach reproductive maturity—oaks and maples, for example—would have a still more difficult time than faster-growing species. Some trees at the extreme southern edges of their ranges would become more vulnerable to pests and diseases that thrive in warmer temperatures. Wildlife that depend on plants and trees would go north too—*if* they can negotiate the obstacle course of highways, cities, and shopping centers.

Forests would become more and more vulnerable with climate change. The old, more stable species would increasingly be replaced by faster-growing "pioneer" species, such as aspen and birch—weedy trees that often invade after a forest fire, a flood, or an avalanche. According to ecologist George Woodwell, we could have several treeless decades after the old regime dies and before new forest systems establish themselves. No optimist on this subject, Woodwell says in *Oceanus* journal that "The sudden destruction of forests by air pollution, now being experienced in Northern and Central Europe and in the Eastern mountains of North America, is but a sample of the destruction that appears in store."

Forests do play a pivotal role in keeping the climate in balance. They are an intricate part of what scientists call the carbon cycle. Through a finely tuned feedback system, nature keeps the amount of carbon in the air, in the sea, and on land relatively constant. Forests absorb carbon dioxide during photosynthesis, store the carbon in their biomass (leaves and wood), and then give off carbon dioxide when they respire at night or as they decay.

"In the process of photosynthesis," Woodwell and Richard Houghton write in *Scientific American*, "terrestrial plants remove about 100 billion tons of carbon from the atmosphere per year, or about 14 percent of the total atmospheric carbon content. An approximately equal amount of carbon is returned to the atmosphere through the processes of plant respiration and decay of organic matter." Because these organic processes are so important in keeping carbon fluxes in balance, changes of a few percentage points in either the photosynthetic or respiratory flux could significantly alter the content of carbon dioxide in the atmosphere.

Rising temperatures could also accelerate respiration and decay—key segments of the carbon cycle. In a stable ecosystem, plant respiration and photosynthesis are roughly in balance. But respiration—the process that releases carbon dioxide—is much more sensitive to increased temperature than is photosynthesis (the process that removes carbon dioxide). As Woodwell points out, when respiration speeds up more than photosynthesis does, trees release more carbon dioxide to the atmosphere than they remove—the situation when seasonal forests drop their leaves in autumn and winter. A temperature-induced increase in respiration could shift the balance and reduce the amount of carbon stored in terrestrial ecosystems, thus adding to the buildup of the greenhouse gases that triggered the initial warming. At the same time, the trees' destruction and decay could send as much as 3 billion tons of carbon in the form of carbon dioxide into the atmosphere.

Loss of trees also leads to the loss of soil, which also stores carbon. Increased respiration by soil bacteria could rapidly raise the biotic contribution to carbon dioxide emissions. Some experts estimate that if temperatures rise by 3 to 9 degrees Fahrenheit, bacteria could release an additional 5 to 10 billion tons of carbon dioxide into the atmosphere.

Countervailing forces could moderate the impact of climate change on the carbon cycle. According to Woodwell and Houghton, "The boreal forest and other coniferous forests may indeed be sufficiently resilient to respond to warming with increased photosynthesis and growth. . . . [But w]hether the carbon taken up by photosynthesis will be stored or simply released through increased respiration remains an open question."

It's possible that the forests may have already accelerated photosynthesis in response to global warming. Forests may be a missing piece in a mystery that is plaguing atmospheric scientists. Researchers have calculated the amount of carbon deposited in the atmosphere by humans—fossil fuel burning, deforestation, and the like—and by natural sources, as well as the amount of carbon taken out by the oceans, forests, and other reservoirs. Discrepancies, however, do exist in their calculations. About a billion tons of carbon known to be emitted into the atmosphere each year seem to be missing in action. Scientists at Oak Ridge National Laboratory speculate that this missing carbon is being taken up through rapid photosynthesis by vegetation and plankton. The increased uptake—the carbon dioxide "fertilization effect"—is believed to be triggered by the extra atmospheric carbon dioxide itself. If so, trees are more important in countering the atmospheric carbon dioxide buildup and deforestation poses a greater threat than previously believed. Unfortunately, scientists have not yet found any ecosystems in which accelerated growth appears to be occurring measurably. Deforestation remains something of a wild card in the carbon cycle. Estimates of the carbon being released through deforestation range from 0.5 to 2.5 billion tons. Because trees store carbon, some scientists argue that widespread, extensive tree planting could help offset the global climate change. The Worldwatch Institute figures that for every acre of trees planted, an average of 2.5 tons of carbon would be "fixed" each year—that is, kept out of the atmosphere where it's a menace. For perspective, World Resources Institute estimates that it would take up to 3 billion acres of trees to offset just today's carbon emissions.

By any reckoning, carbon fixation by plants, trees, and oceanic microorganisms is a mainstay of the global climate regime. If we destroy the biota or reduce its productivity, we are bound to accelerate the processes leading to global climate change.

AND WHAT ABOUT US?—THE HUMAN TOLL

It's unlikely that climate change, at least in the immediate future, will push the comfort zone, within which humans can live and thrive, beyond the limits of tolerance—unless some kind of catastrophic run-away greenhouse effect is set off, and no one expects that. But greater heat has never been equated with good health, and the warming trend, along with the loss of Earth's protective ozone shield due to emissions of many of the same trace gases, could trigger a variety of human ailments—including lung diseases, tropical diseases, and heat-related strokes.

Extra heat alone will be enough to kill some people. Heat waves often produce "excess deaths" to a population. During a heat wave across the United States in July 1980, deaths rose by more than 50 percent, and one study showed that a heat wave of 110 degrees Fahrenheit could increase heat-related deaths in New York City by ten times. In 1980 in Memphis, Tennessee, for example, temperatures stayed above the mean for 27 days; there were 83 heat-related deaths, compared to none during the same month in earlier years. "The poor, the elderly, and inner-city residents, especially those with previous medical conditions, bore the brunt of the heat wave, " write Devra Lee Davis, Victor Miller, and James J. Reisa in a report for the National Research Council.

During the 1988 heat waves, a combination of heat, drought, and floods killed more than 10,000 people in China, and 1,200 died in the 1987 heat wave in Athens. In Chicago, which was hit by a record string of daily highs over 100 degrees Fahrenheit in 1988, 39 heat-related deaths were recorded at the end of August; Missouri registered 30 deaths and 572 cases of illness. Not surprisingly, many of the casualties were old, poor, or both, according to *Time* magazine. But two men in their twenties, who were running in races in New York City, also died.

The number of heat-related deaths may be even higher than pre-viously thought. Heat attacks and strokes account for about 90 percent of all deaths due to heat, says Moulton Avery, executive director for the Center for Environmental Physiology, a Washington nonprofit institute that studies how climate affects the body. Yet, most are not reported as

THE WEATHER FORECAST FOR 2050

Powerful computer models can give climatologists a glimpse of future weather changes that the greenhouse effect might bring. Current models can only partly mimic Earth's many systems and cycles, but climate researchers have predicted that global temperatures could rise as much as 8 degrees Fahrenheit by the middle of the next century. Even a change of only 4 degrees would affect weather appreciably, according to a NASA computer. The model's forecast for 2050 is "hot and dry" for many U.S. cities.

City	Days over 90° F		Days over 100° F	
	Today	2050	Today	2050
Washington, DC	36	87	1	12
Omaha, NE	37	86	3	21
New York, NY	15	48	0	4
Chicago, IL	16	56	0	6
Denver, CO	33	86	0	16
Los Angeles, CA	5	27	1	4
Memphis, TN	65	145	4	42
Dallas, TX	100	162	19	78

heat related, and so heat's toll is underestimated. "If a tornado rips through your town, you know how many died within about three days," says Avery. If a building falls down on people, "you can dig 'em out. You can fly in the governor and take pictures of him declaring it a disaster area. A heat wave is much more insidious. The system for counting the number of fatalities isn't equal to the task."

Avery calls this oversight "a huge public health gap" and estimates that the number of strokes and fatalities linked to heat may be underestimated by a factor of ten. For example, in the 1980 heat wave, only about 1,200 heat-related deaths were reported, says Avery, when about ten times as many may have occurred. How many heat-related deaths occurred in the great national heat wave of 1988 isn't yet known, but Avery estimates the figure at "not less than 10,000 and perhaps as many as 15,000." "These are the largest natural disasters we have, but because there is no property damage there's no attention to it." He also says, "it doesn't present a good photo opportunity . . . and we still do not have a national system for tracking and evaluating these disasters before they occur."

Air quality can also deteriorate as temperatures rise. Hot stagnant air is a perfect incubator for ground-level ozone, the main component in smog, which not only retards growth in trees and vegetation but also damages human lungs and has been linked to increased rates of infection in children. In the 1988 heat wave, U.S. cities recorded the highest levels of ozone in the air in decades, says National Academy of Sciences epidemiologist Devra Lee Davis. "Sweden is now talking about banning cars because ozone levels got so bad that summer," says Davis. She anticipates that cities in this country will also issue restrictions on cars as people are increasingly afflicted by respiratory diseases.

David Bates, former dean of Vancouver Medical School and now an official on the task force on asthma at the National Institutes of Health, notes that exposure to ground-level ozone at levels above the EPA standard of 0.12 parts per million has been proven particularly dangerous for the 5 percent of the population that is asthmatic. A 1982 study of Los Angeles asthmatics found that as ozone levels went up, asthma problems worsened. "Asthma hospital admissions in the United States and Canada have gone up by about 50 percent," Bates notes. "And we don't know why." Researchers suspect that the sudden jump may be connected to increased smog, acid aerosols in the air, or the effects of the two in combination.

Hotter air could also have another nasty effect on air pollution, which is to increase the speed at which sulfur dioxide from power plants turns into sulfuric acid. "And that will most certainly have adverse effects on respiratory health," says Bates, because sulfur is toxic to the respiratory tract. People with respiratory diseases—victims of emphysema and asthma—are especially vulnerable to heat. But with the extra pollution expected, the greenhouse effect could threaten the health or comfort of many others.

As the greenhouse gases and chlorofluorocarbons accumulate in the atmosphere, other problems will be set off or aggravated. As chlorofluorocarbons deplete ozone in the stratosphere, more ultraviolet light will reach Earth's surface. Because ultraviolet radiation is a major cause of skin cancer, greater exposure could thus increase the incidence of skin cancer. According to Adrian Kornhauser, chief of dermal and ocular toxicology for the Federal Food and Drug Administration, a loss of just 2 percent of the ozone layer—which is damaged by some greenhouse gases—would lead to 5- to 10-percent increases in various skin cancers, such as basal and squamous cell cancers and melanoma. EPA estimates that for each 1 percent loss, there could be an additional 570 cancer deaths per year. Over the next century, a 3- to 5-percent loss of ozone in the stratosphere could lead to more than 33,000 cancer deaths in the United States alone.

Ultraviolet rays also seem to suppress the immune system, says Edward De Fabo, a photobiologist at the George Washington Medical Center in Washington, D.C. His research indicates that ultraviolet light apparently turns on so-called suppressor T cells in the skin. These cells counteract "attacker" cells in the immune system that otherwise move in on foreign material. T cells also dampen the skin's natural defenses against cancer cells, says De Fabo. "They lower the body's ability to attack skin cancer and other skin-associated diseases such as herpes and leishmaniasis (an infection thought to be spread by tropical sandflies)," says De Fabo.

These T cells might even be playing a role in AIDS and in lupus erythematosus, another autoimmune disease, De Fabo says, but the effect of increased ultraviolet radiation on these diseases isn't known yet. "What we do know is that this antenna on people's skin could be regulating things we're not aware of," he says. "But we don't know what happens if we stress this mechanism too much."

Eyes too can take only a limited amount of ultraviolet light. The eye's lens filters out those rays to protect the cornea. But, according to Kornhauser, increased exposure can harden the lens and lead to the development of cataracts. "More than one million cataract operations are performed annually in the United States," Kornhauser reports. "This number could significantly increase as we continue to face higher UV levels." Because ordinary sunglasses don't protect the eyes adequately, he recommends special lenses now on the market that block out up to 98 percent of ultraviolet light.

Higher temperature and greater rainfall could also create ideal conditions for a whole host of infectious diseases to spread, through insects, parasites, or food. In developing countries, where these diseases are already exacting a sorry toll, temperature rises could be tragic. "You'll find more infectious disease in the Third World, especially with immune suppression coming from ozone depletion," says Davis. "Then you add common air pollutants, plus malnutrition in these countries, [and] you have a situation where more people are more vulnerable to major epidemics of infectious diseases like typhus and malaria."

"With warming climate, changes in soil moisture and microbial communities may well expand the ranges for exposure of humans to tropical infectious agents," write Davis, Miller, and Reisa. We could be opening the gate to insect-borne infections, including dengue, yellow fever, and malaria, that flourish in tropical climates. Food-borne diseases thrive in the tropics, though poor sanitation and living conditions are most often to blame for these. "Depending on how hot it gets," Davis warns, "we could easily see more mosquitos, sand flies, tsetse flies and other carriers of disease invading temperate regions."

Epidemiologists are quick to stress that climate isn't the sole determinant of disease. Such social and economic factors as proper sanitation and medical treatment obviously govern the likelihood of catching and treating many ailments. For example, William Kellogg and Robert Schware note that malaria was conquered with social weapons in the United States—in fact, almost one-fourth of all Americans live in areas once infected with the disease. But such defenses—public education, sanitation, immunization, and early treatment—are expensive and aren't available everywhere. Tropical diseases have received comparatively few research dollars (considering how many people they debilitate and kill), so a new arsenal of drugs, treatments, and frontline doctors won't be waiting when new outbreaks occur.

As the Lyme tick scare of 1989 showed, diseases can lurk in unexpected places and wreak tremendous distress and fear. Much more epidemiological research into the effects of climate change on public health is needed, researchers say, to prepare us for the worst.

Achieving Climate Stability

Within the limits of the physical laws of
nature, we are still masters of our individual
and collective destiny, for good or ill.

E. F. SCHUMACHER
Small Is Beautiful: Economics as if People Mattered

If government takes just one step to defend society from future global warming, it should be to phase out swiftly the use of the most dangerous CFCs—a move that would also help keep the ozone hole from growing. But one step isn't enough. Truly rising to the occasion will be at least as challenging as landing a man on the moon or rebuilding Europe after World War II. And what we do in the next 30 years, the blink of an eye in geological time, will make all the difference.

Exactly how many steps are needed to stabilize Earth's climate depends on how big they are and when they are taken. But the eight essential moves set forth here form a sound basis for a national greenhouse strategy and provide a framework for the personal actions described in the next chapter.

GETTING MORE WORK OUT OF ENERGY

On a PBS television program aired in 1989, MIT physicist Philip Morrison claimed that the greatest achievement of nineteenth-century physics was defining the concept of energy as the ability to do work. Without this concept, Einstein's conception of the universe would never have come to be. At a more practical level, achieving far greater energy efficiency—that is, doing more useful work with each barrel of oil or pound of coal—may be just as important for the next century. Certainly, getting more out of current energy supplies is the fastest and most cost-effective way to reduce greenhouse gas emissions and to offset the potential impacts of future economic growth. Yet, as Sen. Timothy Wirth of Colorado notes, "Efficiency has been the forgotten stepchild of

United States energy policy." Indeed, "abused child" might be the better analogy.

Experience since the mid–1970s in the United States, Western Europe, Japan, Korea, Taiwan, and many other countries has proven that the economy doesn't have to consume more energy to grow steadily, as experts once thought. As Wirth points out, thanks to engineering improvements driven by high energy prices and to changes in what industry uses energy for, the United States boosted economic output by 36 percent between 1973 and 1986 while keeping energy use constant. Except in industry, no further gains in the economic efficiency of energy use have been made since, mostly because oil prices have been so low. But at least we now know that further progress on this front requires no engineering miracles, and there is no evidence to suggest that any country has yet taken efficiency as far as it can.

Opportunities to improve energy efficiency and cut energy costs exist today in almost all countries. Damp, drafty buildings in Great Britain, "muscle cars" zooming down motorways without speed limits in the Federal Republic of Germany, cookstoves burning chunks of dirty coal in China, industrial facilities leaking steam in the Soviet Union—the list goes on and on.

Sprawling and geographically diverse, the United States probably won't ever become as energy efficient as some smaller countries like West Germany, Sweden, and Japan will. Still, it will be a long time before most countries hit physical or economic limits to conservation. Our country is only half as energy efficient as most Western European countries, but it's far ahead of the Soviet Union and most of Eastern Europe and twice as efficient as China.

But national averages tell only part of the story, and industry's record is mixed. The United States uses about 25 percent less energy to produce a ton of paper than many European countries do. But it uses 50 percent *more* energy, on average, to make a ton of crude steel than Japan does and twice as much energy, on average, to make a ton of cement than West Germany. To bring U.S. industry up to international energy standards—or to set new ones—government might set performance standards for new manufacturing facilities, as it does now for appliances and automobiles. Or it could tax energy use and offer tax credits for investing in new and more efficient manufacturing plants.

Transportation

In transportation, the greatest gains can be made on the road and in the air. New jet engines and composite materials hold out the promise that advanced aircraft can fly longer distances on half the fuel. New graphite fiber composites developed for the "Stealth" aircraft are lighter and stronger than the aluminum skins used in most aircraft bodies. Advanced turbofan engines—machines that look more like old-fashioned propeller engines than modern jets—provide more thrust and go up to twice as far as conventional jets on a gallon of fuel. Although the low fuel prices of the 1980s dampened investors' interest in such ultra-efficient airplanes, controlling the greenhouse problem should be reason enough to introduce them over the next 15 years. To give industry an incentive, government could underwrite the research and development costs and use the new planes as military air transports, as they have for other aircraft in the past, and then give manufacturers incentives to build more of them in many countries through joint ventures and under licensing agreements.

On the road, new car designs such as the VW Auto 2000, the Renault EVE, the Toyota XRV, and the Volvo LCP 2000 are proof positive that cars getting more than 70 miles per gallon of gas can provide safe, comfortable, and efficient transportation for four or five passengers. Because the average new U.S. automobile gets about 26 miles per gallon and annually releases more than its own weight (nearly five tons) of carbon dioxide into the atmosphere, the need to improve vehicle efficiency is obvious. About 10 million new cars are bought in the United States each year, while much smaller numbers of the older and less efficient cars are retired. For this reason, the U.S. fleet's average economy is improving more slowly than the corporate average fuel economy (CAFE) standard would suggest.

But the situation in some ways is even worse than these facts make it seem. As the number of cars on the roads increases, we all feel the increase in congestion and traffic delays. Gridlock made the cover of *Time* magazine in 1988, and it is a daily fact of life in many U.S. cities. When cars are idling along in Los Angeles traffic at average speeds of 13 miles per hour, they aren't nearly as efficient as new vehicles driven under Environmental Protection Agency (EPA) laboratory conditions. To make the fleet more efficient in the real world on real roads, we must improve not only vehicle designs but our urban landscape and traffic flow. We can control the flow of traffic with computer-controlled traffic signals, but it may be even more important to reduce the flow by encouraging companies to adopt flextime work schedules and substitute computer communication for some travel.

Gas-guzzling cars and trucks aren't unique to the United States. Today's global fleet of about 350 million light vehicles could expand to 1 billion in just 30 years. Most of this growth will be in developing countries, where many used cars and trucks from the United States, Europe, and Japan have traditionally made up much of the fleet. Because many developing countries have *no* current fuel-efficiency standards, the highest priority should be keeping their burgeoning fleets from rapidly increasing greenhouse gas emissions. The answer? Accelerated development of cheap, safe, fuel-efficient personal transport—including advanced light-rail systems for urban mass transit, new safer motorcycles and mopeds, and commuter bicycles ridden on special protected lanes.

Ultimately, of course, we need to replace carbon-based vehicular fuels with cleaner-burning fuels, including natural gas, alcohols, and hydrogen. Right now, hydrogen made from solar electricity looks particularly promising. It could be available in the Southwest in little more than a decade if government backs its development. This country also needs pleasant, affordable alternatives to the personal automobile, perhaps including the super-sleek, high-speed trains that have won such followings in Europe and Japan. And we need to push Detroit to design gas "sippers" instead of gas guzzlers and push consumers to buy them—perhaps by raising taxes on gasoline and diesel fuel and using some of these revenues to upgrade and maintain our transportation system.

Dozens of other countries, including many in Europe, have already gone much further. (Gasoline costs 3 to 5 times as much in Japan and Italy as it does here, mainly because of taxes.) According to the American Council for an Energy-Efficient Economy, even after rebates to the poor, a 30-cent-per-gallon tax would generate about $40 billion a year!

The Council also recommends raising fuel economy standards for cars and light trucks. If the average car on the road in the year 2000 got 45 miles per gallon, carbon dioxide emissions per vehicle would drop by more than 40 percent. Especially in need of fuel discipline are light trucks and vans. With sales climbing by about 6 percent annually during the 1980s, these vehicles are vigorously marketed to "rugged individualists," even though three-fourths of all the miles driven in most are logged commuting to work and shopping—not exactly high adventure. Soon, new light pickups and trucks may account for half of all gasoline use, says the Council, because, on average, they get only about 18 miles per gallon. A 35-mile-per-gallon fuel standard for these guzzlers would halve their annual carbon dioxide emissions.

Rebates for gas sippers and taxes on gas guzzlers might also help. These measures would not keep manufacturers from building gas hogs for those who can afford them or give Detroit sufficient incentive to compete with Japan and Europe in the design of sleek new cars and trucks that don't empty the consumer's wallet at the pump. But a government rebate of a few hundred or even a thousand dollars per car could do wonders for environmental protection and U.S. fuel security. What a good use for some of the money collected from a higher gas tax!

Buildings and Appliances

Meeting the challenge to squeeze more useful work out of each barrel of oil or kilowatt-hour will require looking beyond the questions of how we get around and how we manufacture new products. It takes us back to where we live—how we build our homes, schools, and offices and how we use the machines that these buildings contain. Residential and commercial buildings consumed about one-third of all primary energy used in the United States in 1987. Much of this energy is dissipated uselessly as waste heat from light fixtures, electrical appliances, and furnaces that are heating our backyards as they heat and cool our homes.

Buildings last much longer, on average, than cars, planes, or industrial equipment—50 to 100 years for buildings compared to 5, 15, or 25 years for other major energy-consuming devices. So making the building stock more efficient will take more time and patience than trying to upgrade the efficiency of the domestic auto fleet, for example. Even then, the gradual replacement of buildings and equipment is not enough. We need to insulate and otherwise weatherize standing buildings and change how we *use* our homes and offices if we want to get the most "bang for the buck" from our new investments.

Fortunately, governments can do a great deal to modernize architecture and get people to change the way they live in it. No quick fixes are in sight, but various combinations of common-sense approaches should do the job. According to Jim MacMahon of Lawrence Berkeley Laboratories, the key policy strategies are innovation, education, motivation, and mandates.

First, innovation. In the hidebound building industry, companies spend a smaller percentage of their own on funds on research than do businesses of the same size in other important energy-using sectors—chemical manufacturing, petroleum refining, primary metals, or paper production. Government can't force the industry to spend more, but it can support the research and development needed to identify and test new building materials, better construction techniques, and more efficient appliances. It can also restructure tax and fiscal policies so that businesses will be tempted to invest in innovation.

Technology exchanges and transfers with other countries would help the U.S. housing industry meet the coming century's needs speedily and in style. The new techniques developed in Sweden for building comfortable, attractive, energy-efficient homes are especially worth emulating. These modular, factory-built homes are almost as airtight as airplanes, and they can be assembled by a small crew at the home site. If the United States followed Sweden's lead, building new houses would require less time and money (so more people could afford them), and growth in demand for residential energy would fall.

Some people think that these Swedish-style homes are so beautiful that the houses would sell themselves here, but, in fact, the construction work force and the home-buying public would need some help breaking with tradition. Governments should support retraining in the building trades to make sure that the coming generation of carpenters learns how to build energy-efficient homes. It should also offer consumers information and crash courses on the relative energy demands of various apartments, homes, offices, and other buildings. Owners should be required to advertise the energy efficiency of any dwelling unit or commercial building put on the market, much as full disclosure of the financial costs of renting or buying a building is now required. The federal government could also require that the energy-consumption rate of all models of buildings be tested in lifelike situations before they can be rented or sold. State governments could keep a register of the energy efficiency of model homes in all new developments so that home buyers could shop comparatively. If real estate agents were required to reveal a building's energy consumption level in sales documents, as they are now required to state the property tax rate, consumers could avoid buying an "energy sieve" and could more easily make their own trade-offs between energy efficiency and other features.

Going one step further, governments could also motivate consumers to make the right decisions. Arthur Rosenfeld and his colleagues at the Lawrence Berkeley Laboratories have proposed charging a hookup fee for public utilities based on the demand that the building would put on the overall system. If builders or sellers had to pay higher fees for buildings that wasted energy or water, both buyers and sellers would soon hop on the energy-conservation bandwagon. A new home with lots of unnecessary gadgets or every labor-saving device known to man might be charged five or ten times as much to hook up to the local sewer and electric lines than would a simpler, average home using only half as much energy. This sliding-scale fee would not rule out plug-in play-things, but it would make the energy extravagant pay a price that reflects the environmental damages of energy use.

Government may have to "meddle" further in the market if the threat of climate change grows. In the United States and Western Europe, it can take years for energy-smart designs to reach major commercial markets. In the United States, it can take 5 to 10 years for the majority of manufacturers to offer the most energy-efficient refrigerator designs. The Appliance Efficiency Standards Act of 1987 sets mandatory per-formance standards for refrigerators, freezers, air conditioners, and furnaces. Legislation being considered would extend the standards to televisions, microwave ovens, washing machines, and automatic clothes dryers. The next step could be applying these standards to electric lighting, building orientation (which can greatly influence ener-gy use), and other building features. According to the American Coun-cil for an Energy-Efficient Economy, even the current performance standards for refrigerators, water heaters, air conditioners, and other appliances will save the United States as much energy by the year 2000 as 22 large electric power plants could produce. The Council also claims that requiring all fluorescent lamps to match the efficiency ratings of the best now on the market and requiring all incandescent lamps to match those of krypton-filled lamps could save consumers almost $2 billion a year by the year 2000 and save as much energy as 2.5 million households use. Shifting to efficient light bulbs pays a double dividend in the United States: ultra-efficient bulbs release less heat energy, so less air conditioning is needed.

Ultimately, standards for energy consumption could be applied to whole buildings. The federal government, the country's largest user of real estate, could refuse to lease or purchase any buildings that don't meet these standards. It could also make government buildings showcases of energy efficiency, not to mention nice places to work and do business.

Utilities

As for supply, the other side of the energy equation, some comparatively new power-generation technologies convert fuel to heat and electricity far more efficiently than traditional technologies do. Cogeneration systems, for example, efficiently convert most of the energy in the fuel into electricity and then use the waste heat to make steam to heat or cool buildings or drive industrial processes. End to end, this system uses about 60 percent of the heat in the fuel, compared to about 33 percent for conventional steam-electric technologies.

Some industries have put cogeneration to work too. Some U.S. paper mills, for example, convert the extra steam from their pulping processes into electricity. According to the American Paper Institute, nearly half the paper mills' electricity needs are met through cogeneration, which saves up to $2.2 billion each year. Some Brazilian alcohol distilleries even sell cogenerated electricity back to utilities, often at a greater profit than they make by selling the alcohol. Further energy savings are possible in Brazil, according to a recent study by the World Bank and the São Paulo electric utility. With a $2-billion investment in energy efficiency, the utility could save itself the $44-billion expense of building 22 large new power plants by the year 2000.

Energy sources that produce *no* carbon dioxide are obviously the first choice. Luckily, many of these are superabundant. Solar, hydro, wind, and wave power all fall under this category of renewable resources. By comparison, fossil fuels take millions of years to form, and once they go up in smoke, they are gone for eons. Energy from renewable resources, as the name suggests, face no such constraints because the power sources—the sun, for example, or the wind—are self-perpetuating. And if the sun stops shining, we'll have more to worry about than running car engines and steel mills.

Renewables are already in use around the world. Wind turbines both large and small are generating significant amounts of electricity in Denmark, the United States (California), Australia, and the United Kingdom. Large, central-station photovoltaic (solar cell) systems that convert sunlight directly to electricity are supplying electricity in California, Texas, Sicily, and Washington, D.C., among other places. In some Indian and Egyptian villages, solar electricity is used for water pumping, lighting, food refrigeration, and power for medical clinics and communications. Solar-thermal electricity and solar-heated water are being widely produced. The Luz Corporation, a U.S.–Israeli company, operates almost 200 megawatts of solar-thermal electric plants with natural gas backup in southern California. Solar water heaters are produced and exported by firms in Israel, Sri Lanka, India, and Australia. In the Federal Republic of Germany, a prototype photovoltaic system produces clean-burning liquid hydrogen to power a modified BMW sedan.

As production costs decline, advanced technologies that harness renewable energy resources contribute more and more to global electricity supplies. In most of the United States, photovoltaic systems and wind machines already produce energy for about the same cost as the most recently built nuclear power plants do, and wind energy is slightly cheaper. If governments squarely back solar research and development during the next decade, these small modular systems could economically provide 10 percent of all electricity within 20 years.

How does nuclear energy fit into the new energy age governments should promote? Although the prospect of global warming has renewed interest in every type of "smokeless technology," nuclear power's future remains uncertain. Commercial plants now produce less than 10 percent of the world's electricity. Within the next 20 years—key decades for responding to the greenhouse problem—this share probably won't grow because plants take several years to plan and license and another 5 to 15 years to build. Nuclear facilities are also expensive to construct —as are some solar facilities. Then, too, the public's fear of mishaps in the wake of the comparatively minor accident at Three Mile Island in 1979 and the disastrous one at Chernobyl in the Soviet Union in 1986 has slowed construction almost everywhere except in France (where everyone who lives near nuclear power plants gets free electricity). Given the alternatives, efficiency continues to look better and better.

One way for governments to streamline electricity generation—and its global 30-percent contribution to carbon dioxide emissions—is to take a new approach to utility regulation. Utilities make money by selling electricity, so what incentive do they have to conserve? One answer is that some new power plants—especially nuclear facilities—cost so much to build that they can't pay for themselves unless utilities significantly raise their rates. But what about oil- or coal-fired power plants that aren't as expensive to build as nuclear plants but contribute greatly to the greenhouse problem? To push *all* utilities into a new energy era, government should reward the forward-looking ones that invest in energy efficiency. Because adding new capacity—especially nuclear capacity—has become so budget-breaking in recent years and because utilities aren't particularly eager to subject their customers to "rate shock," many power companies are quite willing to cooperate. In California, the Public Utilities Commission has experimented with allowing higher rates of return to utility companies that strive to increase energy efficiency. Under this plan, even if it generates and sells fewer kilowatt-hours of electricity, a utility can earn higher profits than it otherwise would, and utility executives get the message that society values efforts to increase energy efficiency and thus reduce polluting emissions.

Government can also help utilities "empower" their customers. Traditionally, utilities invest in equipment that the customer neither owns nor operates. But programs directed at the customer's side of the electricity meter can mean extra savings for both the utility and the customer. In California, state standards for appliance efficiency, coupled with special rebates, nudge consumers in the right direction. In one program, consumers who buy a refrigerator from the utility's list of highly efficient models and allow the utility to dispose of the old model safely get a $50 rebate from the power company. The utility makes money because it saves more than the cost of the rebate by not having to invest in new generating capacity to run a less-efficient refrigerator in each home. The consumer makes a quick $50 and—more important—also enjoys lower monthly bills for as long as the appliance lasts. With one small addition, the program could be made even better. If the utility made sure that the refrigerator was drained of chlorofluorocarbons (CFCs) before it was junked, the CFCs in the old refrigerator would never make it into the atmosphere to contribute to both global warming and ozone depletion. One refrigerator-recycling program in Austin,

Texas, has helped the municipal utility there cut the electricity growth rate by roughly 30 percent for a fraction of the cost of finding new power supplies to meet expanding needs. Energy expert Amory Lovins of the Rocky Mountain Institute calls such highly valuable saved energy "negawatts."

One more important change in regulatory policy will also help defend us against climate change. If utilities were allowed to pass along the environmental costs of energy supply and use to their customers, corporations and individuals would get a more accurate picture of energy consumption's full consequences. Costs that economists call externalities—once defined by Barbara Ward and René Dubos as "the costs an enterprise can cause to others while escaping damage itself"— would no longer be hidden. Historically, such costs have been ignored, and society as a whole ends up subsidizing energy consumers, especially heavy or wasteful consumers. In the past, these subsidies have made energy look cheaper than it is. But if polluters were forced to pay for the damages they cause, they would pass them along and people would come to recognize the true, total cost of energy use. People would see when it makes more sense to conserve energy than to buy more of it—to use "negawatts" instead of megawatts. As environmental costs raise the market price of electricity, for instance, consumers and corporations will use it more sparingly, pushing manufacturers to design ultra-efficient appliances and energy systems.

In developing countries, where many people cook with wood and charcoal, the pressing need is to make the combustion of wood and other types of biomass more efficient. (That way, none of the carbon in the fuel is released as methane.) The challenge is to make small stoves that transfer heat very efficiently and to make them simple to produce, easy to use, and inexpensive. Here, the U.S. government's role should be using foreign-aid dollars to help other governments and agencies limit tropical deforestation and introduce more efficient cookstoves.

SWITCHING FUEL AND CONTROLLING EMISSIONS

While we wean ourselves from coal, oil, and natural gas, *which* fossil fuels we burn makes an enormous difference: not all pollute equally. Coal, for instance, produces almost twice the carbon dioxide that natural gas does for the same amount of energy, and one-and-a-half times as much as oil. Synthetic gas made from coal is even worse.

Besides the extra carbon dioxide in the coal itself, the gasification process adds carbon dioxide to the atmosphere, making total emissions almost three times as high as those from burning natural gas alone—not to mention the accompanying solid and liquid by-products. Clearly, until we build a smokeless energy system, natural gas is the fossil fuel of choice.

One "new" fuel worth considering as part of a defense against greenhouse warming is hydrogen—one of the cleanest-burning fuels available. When electricity is passed through water, the water molecules split into hydrogen and oxygen. The hydrogen gas that results can—much like natural gas—be piped to wherever it is needed to produce heat, power slightly modified vehicles, or manufacture agricultural fertilizers, hydrogenated food oils and fats, and other industrial chemicals.

In the 1970s, engineers and policy makers flirted with the hydrogen option. But this fuel was never widely used because electricity sources needed to make a hydrogen system work remained too expensive. Luckily, recent advances in solar photovoltaic (PV) technology are making hydrogen competitive with the fossil fuels we will be using a decade from now. In the Southwest and the other sunnier regions of the United States, the direct current electricity needed to produce hydrogen could be available, from solar sources, by the turn of the century for about what electricity from coal—or gas-fired power plants—costs, according to Joan Odgen and Robert Williams of Princeton University's Center for Energy and Environmental Studies. At this cost, hydrogen could help societies cope with urban air pollution, acid precipitation, and global warming. What is government's role here? Increasing support for photovoltaic research and development and sponsoring commercial demonstrations of photovoltaic hydrogen systems to make chemical and power companies, utilities, and car manufacturers sit up and take notice.

Another challenge for both policy makers and engineers is controlling methane leaks. What we call natural gas is really about 98 percent methane—a far worse greenhouse gas than carbon dioxide. In energy production, use, and transport, methane is vented directly into the atmosphere. It leaks from coal seams, old oil fields, natural gas pipelines, and fuel-distribution systems. Natural gas lost during extraction and transport of fossil fuels may account for one-tenth of all methane emissions worldwide. In the United States, gas companies lose track of 2 to 3 percent of all the natural gas they extract. Some losses reflect leaks; others, theft and meter malfunctions. In Eastern Europe and developing countries, losses may be six or seven times greater. For government, probably the best way to fight those losses is to reward companies for good pipeline maintenance and to fine them for leakage.

Apart from at-the-source measures for controlling methane and other greenhouse gases, can anything be done once the culprit gases have been released? With a few exceptions, no. Carbon dioxide can be removed from exhaust gases before they are vented into the atmosphere. Indeed, this gas is already extracted from so-called stack gases in a few factories (and then put to work in food production, chemical processes, and some oil-recovery operations). But this remedy is fraught with problems of its own. It is prohibitively expensive. It's untried on a large scale. It decreases the power plant's efficiency. And the whole U.S. market for carbon dioxide (about 20 million tons) is less than 10 percent of the carbon emitted by coal plants alone. Even if carbon dioxide were scrubbed out of the flue gas released by fossil-fueled power plants, permanently disposing of the massive quantities of this greenhouse gas would still be a headache. Burial in the deep ocean or outer space has been proposed by extravagant thinkers, but the expense, impracticality, and environmental consequences of handling billions of tons of gas have raised more eyebrows than expectations. Economists are especially leery of the idea: by one estimate, removing half the flue gas from large electricity plants, building pipelines to the sea, and sinking the gas in the deep ocean would double the cost of electricity; removing 90 percent would more than triple electricity costs!

TURNING TO NATURAL DEFENSES

The potential for cutting greenhouse emissions from agriculture and forestry—which account for up to one-third of all nitrous oxide emissions and substantial fractions of carbon dioxide and methane emissions—is tremendous. Because forests and soils absorb carbon that would otherwise remain in the atmosphere, reforesting land (also enhancing the carbon in soil) helps stabilize atmospheric carbon dioxide concentrations. Just how much carbon can be sponged up this way depends on which forestry practices are used, how long the trees planted take to mature, and how the wood is ultimately used. If rainfall is adequate, a new Douglas fir plantation in Oregon or Washington could sequester millions of tons of carbon for a full century. In the tropics, increasing long-term storage is harder because few forests can withstand the *human* pressures of a century of population growth and poverty. In these regions, local needs for fuel, fodder, poles, and protection can be met by multiuse forestry projects. Even fast-growing trees can help stabilize climate by protecting soils and providing lumber to displace fossil fuels.

As important as forestry efforts are in the battle against climate change, we can't just reforest our way out of the greenhouse problem. According to World Resources Institute's Mark Trexler, "Simply offsetting current carbon dioxide emissions for the next forty to fifty years would probably require planting an area much larger than Australia. Dedicating this much arable land to new forests is out of the question, especially when the initial warming could throw some types of forest ecosystems off balance and further increase carbon emissions." Even if enough suitable land could be found and freed up for forests, human population growth, land-ownership patterns, and other long-standing social concerns could get in the way. Then, too, scientists still aren't sure whether the world's greenery can continuously absorb enough carbon to offset fossil fuel combustion, much less reverse the accumulation of greenhouse gases.

Despite these uncertainties, action has already begun. In 1989, the first "carbon offset" project was launched with considerable fanfare. The equivalent of the 15 million tons of carbon that will be emitted over 40 years from a coal-burning power plant built in Connecticut by Applied Energy Services will be absorbed by an enormous woodlot in a combined forestry and agricultural project in Guatemala. The energy company donated $2 million to this maiden effort, and CARE, the Peace Corps, the U.S. Agency for International Development, and the Guatemalan Forestry Department are also chipping in. Altogether, some 52 million trees will be planted in what the *National Geographic* calls "the most sensible and imaginative program yet conceived to put the industrialized world's money where its mouth is."

Besides supporting more projects like the one in Guatemala, government should develop guidelines for "emissions trading" among new industrial sources of sulfur dioxide, nitrous oxides, and perhaps other greenhouse gases. A new plant that couldn't meet emissions standards, for example, would have to pay other industries to compensate by reducing their releases. Alternatively, the new plant might have to clean or scavenge an equivalent amount of heat-trapping gases from the atmosphere. If a market for such offsets could be developed, direct daily government intervention would not be needed to make this plan work.

Innovation will also go a long way toward controlling greenhouse gases coming from farms and ranches. For instance, just putting a lid on manure piles and feedlots would allow us to capture and use methane as a high-grade fuel, thus turning a problem into a profit-maker. Developing productive rice strains that don't have to be flooded would eliminate methane-producing microbes. Increasing soil productivity by limiting erosion and strategically irrigating growing plants while cutting back on fertilizers would also reduce nitrous oxide emissions, lock carbon in the soil, and reduce the deforestation that occurs when worn-out agricultural plots are abandoned and new lands cleared.

SAYING GOODBYE TO CFCs

The Montreal Protocol is designed to halve the production and use of the most dangerous CFCs in the industrialized countries by the turn of the century. By late 1989, international consensus had strengthened against CFCs: 81 countries signed the Helsinki Declaration to ban CFCs by the year 2000, and this declaration may become an amendment to the Montreal Protocol. But putting pen to paper does not magically whisk CFCs out of the atmosphere. Many of these gases linger in the atmosphere for decades or centuries before natural processes break them down into harmless, simple compounds.

To bring CFC emissions rapidly down to the level set in the Montreal Protocol, government will have to use some combination of carrots and sticks to get industry to comply with the protocol's provisions. An obvious starting point is to encourage industry to find substitutes for CFCs, especially the most dangerous ones. Several formulations of new, safer CFCs (including CFC-22, CFC-142b, and CFC-152a) are now available for some applications, and others (including CFC-134a, CFC-123, and CFC-141b) are being developed. All are less dangerous than the CFC-11 and CFC-12 now commonly used.

It's also possible to produce CFC-based products with greater care. There is great potential for recapturing the agents used in manufacturing some insulation, plastic products, and electronic equipment. Enclosing process lines, aging CFC-based products before shipping them, and

KICKING THE CFC HABIT

A big part of saving Earth's ozone layer is environmental diplomacy—persuading world leaders, environmentalists, and industrialists to ban CFCs. But an equally big part of the solution is finding safe, cheap, and effective alternatives for the more than 2 billion pounds of CFCs produced every year worldwide. Since their invention in 1930 as refrigerants, CFCs have crept into a wide range of industrial products, from Styrofoam cups and fast-food containers to solvents for cleaning electronic circuits and coolants in refrigerators and air conditioners. Whether or not American consumers have heard of CFCs, they have bought and used them.

The E.I. Du Pont de Nemours Company, which produces almost half of the CFCs in the United States, surprised many ozone watchers in 1988 by announcing plans to phase out the harmful chemicals by the turn of the century. Du Pont predicts that acceptable substitutes could be on the market within five years, but would cost from two to five times as much as CFCs. These more expensive replacements are also trickier to manufacture. Currently, CFCs are made from inexpensive materials in a one-step process that yields a high-quality, relatively pure product. Alternative manufacturing methods run at hotter temperatures, adding to the cost, and take several steps to obtain a pure yield. Du Pont is testing a line of likely replacements that include fluorocarbons (CFCs without chlorine) and altered forms of CFCs that break down before they reach the upper ozone layer.

The threat of a CFC ban has spurred many chemists and businesspeople to innovate. One Florida researcher, collaborating with AT&T, found that orange peels might help in the battle against ozone depletion. Citrus rinds contain oils called terpenes that can be used instead of CFCs to clean electronic circuit boards. Unlike CFC solvents, terpenes break down naturally near the ground and don't pose a threat to the ozone layer.

An aerospace company, Cryodynamics, Inc., boasts a space-age refrigerator that uses helium instead of destructive CFCs for cooling. The company is pushing a new line of its "cryodynamic coolers" for household use that have become hot-selling items since news broke on the ozone hole. Recently, the Chinese government ordered a million of the helium iceboxes for domestic use.

(continued)

KICKING THE CFC HABIT   *(continued)*

One unlikely CFC replacement is ammonia, the toxic chemical that spurred the search for CFCs in the first place. The use of ammonia never died out completely—it is routinely used in mobile home refrigerators—and today's safer refrigerator designs make using ammonia less hazardous than it was in refrigeration's early days.

The plastic foam industry—the largest CFC consumer in the United States—may be the hardest hit in the scramble to adapt to a ban. Carbon dioxide and methylene chloride can be used to make some foam products but, inch for inch, foam blown with carbon dioxide can't insulate as well as CFC-blown foam. Home builders looking for cheap ways to meet energy-efficiency standards face a much tougher and more important search for affordable, high-performance alternatives. Some of Du Pont's safer CFCs might do the trick, but these costly replacements are still some years away.

What do market analysts predict? After initial price jumps and disruptions, consumers and industry will adapt to a CFC-free world in as little as 10 years, and new markets for the innovative products and technologies will open up.

---

carefully controlling spray heads can substantially reduce airborne emissions in manufacturing and may also allow industry to recapture the dangerous gases. As for the CFC-based solvents used in the electronics industry—the fastest-growing class of CFCs —conservation techniques are already coming into use. The Hewlett-Packard Corporation, among others, minimizes its solvent losses by enclosing its production lines and passing the captured gases through a bed of absorbent carbon.

Governments can speed the introduction of "loss-control" technologies by imposing industrial health and safety standards. If implemented by the U.S. Occupational Health and Safety Administration and its counterparts abroad, such standards would limit ambient levels of CFCs in and around factories. In such products as foams, refrigerators, and fire extinguishers, leak testing can be required prior to shipment. Combined with fines and other penalties for releasing pollutants, these measures could quickly reduce CFC losses during manufacturing.

Besides promoting efficiency, governments can also encourage industry to recycle these compounds. Congress is now considering placing a deposit fee on CFCs, similar to that imposed on glass containers in some cities and states. Such a fee would increase the value of "used" CFCs and encourage recapture. In addition, our government could establish common maintenance and service procedures. Garages and auto dealers servicing car air conditioners could be required to recapture and recycle the used refrigerant. But some institutional obstacles to recycling must first be swept away. For one thing, the Department of Transportation would have to drop its ban on refilling the containers used to ship CFCs.

The bad news about alternatives to CFCs is price. Suitable substitutes cost from one-and-a-half to five times more than the more dangerous, traditional CFCs, which are currently mass-produced from cheap feedstocks. To manufacture alternatives, new factories must be built, new tests run for stability and toxicity, and new service people trained in their installation and use. For at least a few years, these costs are likely to keep prices for substitutes higher than those for the more common and dangerous CFCs they replace. To help, government should organize public-private partnerships to accelerate research and development, as it has for the microelectronics industry.

Another worry is that the substitutes themselves may be greenhouse gases or ozone depleters. And we don't yet know whether these compounds will spur chemical reactions that create air pollution in the troposphere. Considering that we got burned with the original CFCs— once considered totally safe—it makes sense to be cautious with these. As new CFC substitutes are developed and tested, government needs to make sure that these compounds' degradation in the atmosphere is completely understood. In many cases, the new CFCs are less dangerous in that they break down faster than the traditional formulations they replace. Unfortunately, the same molecule that triggers their breakdown (the hydroxyl radical, or OH-) is the same one we are depending on to remove methane from the atmosphere, so local "shortages" might occur as the new, improved CFCs are produced. To reduce this risk, government needs to regulate automobile emissions and make biomass use more efficient.

MAKING ECONOMICS ENVIRONMENTALLY HONEST

No scientific breakthroughs are needed to bring civilization's energy future in line with Earth's needs by 2025. But some economic reconceptualizing and perhaps some new regulatory laws may be required.

First, the environmental costs of energy production and use must be built into energy prices. Right now, the true costs are much higher than the prices we pay. A real, dig-in-our-heels commitment to "getting the prices right" means scrapping subsidies that distort market prices and that encourage energy use. A carbon tax and consumption taxes on energy use would be steps in the right direction. So would eliminating a whole host of subsidies that increase demand for fossil fuels: tax deductions for the extraction and transport of energy, subsidized fuel exploration and production, direct cash payments, interest rate subsidies, and so on.

For the record, the United States is not the only country beguiled by energy prices that invite resource waste and environmental trouble. Governments all over the world manipulate price controls, exchange-rate differences, and direct subsidies to keep the prices consumers pay for energy below what they would be in a free market. According to resource economist Mark Kosmo, the retail price of natural gas in developing countries ranges from 30 to 62 percent less than the international market price. As for oil, both producing and importing countries subsidize its use. In Bangladesh, a bitterly poor country, the fertilizer and electric power industries get such a deep discount on natural gas (30 percent on newly developed supplies) that they gobble up almost three-fourths of the country's total supply. Chinese consumers get an almost unbelievable deal, paying only about one-fourth of the international market price for coal. Of course, *somebody* pays: in 1983, subsidies to coal users cost mainland China's government $10.5 billion, almost 4 percent of its gross national product.

Unfortunately, these subsidies all encourage fossil fuel use. Different countries back different fuels to different extents. The key point to remember is that all such subsidies make inefficient and environmentally damaging fuels look artificially cheap to the consumer. The other side of the coin is that it pays to use energy efficiently. As Rep. Claudine Schneider notes, "Since 1973, energy efficiency gains have displaced the equivalent of 14 million barrels of oil per day, saving

Americans more than $150 billion per year. . . . . Foreign oil imports are less than half of what they would have been, lowering the annual trade deficit by more than $50 billion. Efficiency gains worldwide have been instrumental in spawning a global oil glut, collapsing world oil prices, curtailing OPEC's power, and reducing inflation rates fanned by high energy costs." All these gains are in addition to the environmental benefit: carbon dioxide emissions are 50 percent lower than they would otherwise have been.

FILLING THE KNOWLEDGE GAPS

If we had no satellites, no worldwide network of weather-monitoring stations, no research laboratories, no interdisciplinary institutes, no super-computers, and no high-tech ice core drills, we might still be blissfully ignorant of the growing threat of climate change. On the other hand, science and technology have not given us complete and perfect knowledge. Important gaps remain in understanding how our finely tuned planetary systems work together, and scientists studying climate change have their work cut out for them.

Scientists know a lot about the carbon cycle, but they don't yet know how several important pieces of the puzzle fit together. Little is currently known about the process that governs the uptake of heat and carbon by the oceans, for example. The role played by the great oceanic currents (the Gulf Stream, for example) and the forces that affect the vertical distribution of heat are also still partially veiled in mystery.

Nor is everything known about how methane figures into climate change. Why is it increasing in the atmosphere by 1 percent a year? How much do methane leaks add up worldwide? Do these leaks negate the benefit of reducing carbon dioxide emissions by switching from coal to natural gas? And on a wide scale, how can we trap and use the methane released from coal seams, depleted oil fields, and natural gas pipelines?

Remaining questions about nitrous oxide are just as pressing. Some research suggests that burning coal and fuel oil gives off this "laughing gas," but burning other fuels does not. Other recent measurements show no significant difference. Which claim is right? Additional research could determine whether advanced combustion technologies could reduce emissions of nitrous oxide (and carbon dioxide) without substantially increasing energy costs. As mentioned in Chapter 2, other evidence suggests that the soil bacteria that digest certain fertilizers release this gas. Research carried out so far indicates that the most popular fertilizers in the United States, those based on anhydrous ammonia, cause far more extensive atmospheric problems than do the urea-based fertilizers commonly used in mainland China and other developing countries. Researchers now need to determine whether something can be done to limit releases from the fertilizers used here or whether we, instead, need to look for new types of fertilizers and new, safer ways to apply them. Or consider the forest/nitrous oxide connection: in 1986, two U.S. foresters reported that clear-cutting increased nitrous oxide emissions from forest areas by as much as 100 percent. If we knew exactly how forests recycle nitrogen, we could reduce nitrous oxide emissions and manage forests better.

Some aspects of climate dynamics are also still beyond our grasp. How clouds form and what determines their behavior still elude atmospheric researchers. Scientists' understanding of how the oceans influence climate is still primitive compared to their understanding of land-based ecosystems and weather systems. As for simultaneous, worldwide measurements—a critical ingredient in climate studies—Carl Wunsch of MIT says that oceanography lags 20 years behind meteorology, partly because ships fitted with sophisticated measuring equipment cost upwards of $30,000 per day to run, and partly because the oceans cover six-sevenths of the planet, making comprehensive readings at regular intervals hard to arrange. Yet, this information is needed to improve the computer models that scientists are constructing to get a fix, first, on whether climate can be accurately predicted and, if so, on how.

Then there is the vexing but fascinating business of feedback processes. As the many examples of feedback loops mentioned in this book make clear, small additional perturbations in one part of the biosphere may provoke disproportionately large responses in others or amplify the original disturbance. For scientists, the challenge now is identifying what those responses are and estimating when we might experience them. Will, for example, a rise of 4 degrees Fahrenheit provoke, as some scientists suggest, an additional release of up to 16.5 billion tons of carbon from soil microorganisms, effectively doubling the release from fossil fuels?

Scientists today have only a primitive understanding of how the biological pump moves carbon from the warm, surface layer of the ocean into the cold, deep layers below. Will increased exposure to ultraviolet light due to stratospheric ozone depletion kill off the tiny one-celled plants that help the oceans remove carbon dioxide from the atmosphere? Or, as recent investigations suggest, could methane trapped in Arctic ice be especially susceptible to a global warming? And will a warming of 6 to 8 degrees Fahrenheit at the poles release vast quantities of methane trapped under the tundra, thus rapidly accelerating global warming? Only through a long-term commitment to basic earth science research can scientists puzzle out the answers to these difficult questions.

None of these answers comes cheaply, and right now global-change research and monitoring are underfunded. Improved monitoring from space, through remote sensing by satellites, is essential. Fortunately, new instruments to measure sea-surface temperature (a useful early indicator of climate change), deforestation, and patterns of cloud formation are to be developed as part of the International Geosphere-Biosphere Programme beginning in 1990. Only through such broad-based cooperative efforts can the human race learn enough about our fragile blue planet to keep its land surface green.

More support for research is needed in other advanced industrial countries, too. Japanese, Soviet, and European scientists are already planning parallel and complementary studies. Because the scale of the problems is so large, more such international cooperation will be our only hope to yield results in time to do us any good. For expediency's sake, experts from various countries must explore separately the areas they know best and then coordinate the results with others to forge new understandings. The recent cooperation of Soviet, Swiss, and French researchers in drilling and analyzing the ancient ice cores that are climate's historical records stands as a model here.

Equally high on the science agenda should be research into ways to make particularly vulnerable groups of people and geographic areas more resilient in unavoidable climate changes. More research is needed on agricultural practices and new crop varieties to help agriculture withstand climate shifts and increasingly mercurial weather. New strains of familiar crops must be developed that can produce high yields even when the weather gets harsh or chancy and water is in short supply. New seeds and varieties of trees that can withstand exposure to high temperatures, higher levels of ultraviolet radiation, and increased air pollution need to be bred.

Applied research on coastal zones also figures centrally in any defense against rapid climate change. How can we defend coastal zones, and when should we abandon regions that the sea will inevitably claim? And how can we protect local resources, including fragile ecosystems and spawning grounds, in the face of rising tides?

More policy research is also needed. As the scenarios sketched in this book suggest, finding the most rapid, complete, economical, and socially nondisruptive ways to reduce greenhouse gas emissions should be the aim. What the public thinks about all these options needs to be investigated and made known to world leaders *before* sweeping life style changes are legislated. Public hearings would help, and so would open discussions in international forums. Because rapid climate change will affect all aspects of human life, it should be on the agendas of organizations ranging from the World Health Organization to the Food and Agriculture Organization, the U.N. Environment Programme, and the World Meteorological Organization.

The United States and other nations now have experience with reducing pollution and promoting energy efficiency. Research on the relative pros and cons of tax incentives, gasoline and oil-import taxes, pollution taxes, trade-offs, penalties, regulations, and negotiations is plentiful. What is needed now is to apply each of these ideas—and such new imaginative suggestions as the debt-for-tropical-forest swap, which would forgive Third World debt in exchange for protecting key tropical forest areas in perpetuity—to specific greenhouse problems to determine which work best under what circumstances. Such policy research is perhaps the only way to get free of the ideological biases that engulf most policy debates.

To achieve those ambitious goals, we should throw support behind the National Aeronautics and Space Administration (NASA) "Mission to Planet Earth." Research conducted under this banner will provide us with a clearer vision of the greenhouse problem and its relationship to other international environmental issues. It could also help identify, develop, and implement cost-effective preventive policies and adaptive responses. And it will certainly spur unprecedented international cooperation in technology development.

BUILDING INTERNATIONAL COOPERATION

Just as citizens can't combat climate change alone, national governments can't ignore the international dimensions of rapid climate change. The reasons are many. Emissions sources dot the globe. Greenhouse gases have long residence times in the atmosphere and can travel continental distances. Climate change defies the cartographer's neat boundaries. And perhaps no other environmental problem is so pervasive.

Currently, the world's industrialized countries contribute over 80 percent of the emissions that commit the planet to future global warming. During the next decades, greenhouse gas emissions from developing countries are sure to increase. With the responsibility thus divided, the first item of business must be to reduce emissions in industrialized countries substantially. The equally important longer-term challenge is to help developing countries obtain the highly energy-efficient technologies that don't emit more pollution than Earth and its inhabitants can handle. But how?

Luckily, the groundwork has been laid, and opportunities for enhancing cooperation are many and clear. Following the long, hot summer of 1988 and the unusually warm winter of 1988/89, concern about the risks of rapid climate change percolated to the top levels of national governments. Prime Ministers Mulroney of Canada and Thatcher of the United Kingdom both committed themselves to reduce greenhouse gas emissions. Prime Minister Brundtland of Norway called for a new Law of the Atmosphere to protect Earth's shared resources. In May 1989, President Bush announced that the United States would host a workshop to explore the opportunities for a global climate convention. Buoyed by the political success of the Montreal Protocol for protecting the ozone layer, leaders from across the political spectrum have entered the international dialogue on climate change with high hopes. In July 1989, the environment was the centerpiece of an international summit held in Paris. At that gathering, President Bush and other world leaders called for limits on emissions of carbon dioxide and other greenhouse gases, along with decisive cooperative action to protect the environment.

Right now, the most important forum for this dialogue is the Intergovernmental Panel on Climate Change formed in 1988 by the U.N. Environment Programme and the World Health Organization. This panel will give all nations a current scientific reading of the greenhouse problem, evaluate policy responses to it, and propose a treaty or some other legal mechanism for reducing greenhouse gas emissions. One politically important question is whether the panel—due to announce its recommendations in late 1990—will propose basing future emissions reductions on countries' past contributions, on current economic activity, or on estimates of future emissions. If emissions are reduced by a fixed percentage, such countries as Japan, which have already achieved relatively low levels of per capita emissions, will seek credit for gains already made. If future reductions are based on the level of cumulative emissions in the past, the United States will object, arguing that "what's past is past" and that we should move forward on equal terms. If future reductions in the industrialized nations are based on the economic wealth of countries or regions, the soon-to-be-integrated European Economic Community is likely to perceive such an allocation formula as a penalty for economic cooperation. Clearly, none of these schemes is perfect, and necessary compromise will come only with a spirit of cooperation born of political courage.

A key issue in all international negotiations will be how the richer countries can promote the development of lean and clean technologies in poorer ones. Technology transfer, joint ventures, and financial risk sharing are all needed. If handled correctly, technological advances in industrialized nations could make a dent not only in the greenhouse gas buildup but also in poverty itself. If provided with more efficient steel-making technology, for example, developing countries might be able to resist the temptation to make steel with charcoal. If they did, carbon dioxide emissions would drop and so would the pressure to decimate Third World forests.

Cooperative technology development should emphasize alternative energy technologies. There is a long and productive history of international cooperation on expensive, new energy technologies, especially nuclear fusion. Applying the same approach to the development of energy-efficient technologies and renewables—solar, wind, hydropower, and biomass energy systems—could yield transforming results. Faced with the current threats to economic stability and environmental quality, the political and corporate leaders of the rich, industrialized countries must overcome a history of narrowly self-serving competition. We must all recognize the common threat of environmental degradation and cooperate on the development of low-risk, low-waste technologies to minimize these risks.

The history of technology development and transfer also contains some bitter lessons that we don't have time to relearn. One is that such efforts will almost certainly fail if the people who are to use the technology don't have a strong say in how it is developed. Another is that the "turnkey" approach doesn't always work. Although a car made in one country may work just as well in any other, a power-generating station might not: it might be incompatible with the power-delivery system, local needs for power, or available fuel supplies. Experience has also shown that a technology's success has a lot to do with who owns it, so developing countries should at least share in the profits of any new energy technology introduced.

With these pitfalls in mind, the U.S. Agency for International Development (AID) and the industrialized countries' other aid agencies could refocus foreign aid—especially funds earmarked for energy—to set new precedents for technology transfer. Large, conventional supply projects should be passed over in favor of many small improvements in how energy is supplied and used. Where new, cleaner technologies are available, aid donors must make sure that they meet local communities' real needs.

Through licensing agreements, joint ventures, and public-private partnerships, governments and aid agencies can push private businesses in developing countries to develop these promising new technologies. One recent effort shows great promise. AID is cooperating with India's government to fund joint ventures between U.S. and Indian firms working together to adapt high-efficiency technologies developed in the United States for manufacture and use in India. The two governments shoulder part of the economic risk of technology development; the private firms provide the labor and capital for the manufacturing.

The threat of global warming also gives our government one more reason to reassess the terms of trade we give Third World countries. We can't, on the one hand, insist upon profound structural changes in these economies and, on the other, cling to protectionist trade agreements and laws. If the markets of the industrialized world aren't soon opened to the products of the newly industrializing countries, these nations won't have enough foreign exchange to meet daily survival needs, much less to retool industry, transportation, and agriculture. And if we don't liberalize trade, the developing countries will have no choice but to buy the cheapest (and least efficient) technologies on the market and to continue to pay off their debts by clear-cutting and burning down their forests—major economic assets and future sources of wealth.

If the prospect of climate change challenges U.S. foreign policy, especially that toward the developing world, perhaps the consolation is that that policy needs revamping anyway. As Jessica Tuchman Mathews of World Resources Institute puts it in *Foreign Affairs*, "Environmental strains that transcend national borders are already beginning to break down the sacred boundaries of national sovereignty . . . and the once-sharp dividing line between foreign and domestic policy is [now] blurred."

This country also needs to show a level of compassion and cooperation that has made only fleeting appearances during "photogenic" famines and natural disasters in recent years. Longtime international environmentalists Barbara Ward and René Dubos said it best nearly two decades ago: there's "only one Earth"—ours.

AVOIDING "STANDING ROOM ONLY"

The rate of future greenhouse gas emissions will be strongly affected by the number of people living on the planet in the next hundred years. There are no firm and immutable limits to the number of earthlings Earth can support. The relationships among population growth, resource use, and environmental quality just can't be boiled down to simple formulas. But, in general, the more people on the planet, the more energy and materials needed to feed, house, clothe, and employ them, and the more exhaust gases and other pollutants released by that energy.

The historical record is clear, even if our vision of the future is clouded. The world's human population has grown more in the last century than in any other comparable period. It took 10,000 years or so for Earth's population to reach a billion by the early part of the nineteenth century. A little more than a century later, global population had about doubled, approaching 2 billion. A third billion was reached by 1960 and a fourth by 1974. By the end of 1987, we had more than 5 billion neighbors, and now the total is increasing by about 100 million per year. Although growth rates have declined somewhat since they peaked in the 1960s, there is still considerable momentum for future growth.

How fast global population will increase and where it will stabilize are highly uncertain. According to "mid-range" U.N. projections made in the early 1980s, global population was to stabilize at about 10.2 billion inhabitants by 2100. But in 1989, that estimate was revised upward to 14 billion, mostly because fertility declines throughout the decade turned out to be slower than had been anticipated.

What will happen in developing countries, where most of the growth is expected to occur? Rapid population growth will strain developing countries' ability to provide employment, adequate nutrition, public services, health care, and protection of environmental quality. Just feeding all those hungry new mouths may require substantially increasing croplands, applying more fertilizer, and increasing irrigation. Unless great care is taken, such programs will add to chemical pollution, soil erosion, deforestation, and salinization. And as demand for nonrenewable resources and fuels goes up, the quality of air and water supplies is likely to deteriorate.

We need not—as some have suggested—depend on the age-old scourges of war, famine, and pestilence to get us out of this mess. Nor should we blame the world's problems on the growing populations of the developing world. But if we don't slow global population growth, reducing the risks of rapid climate change and avoiding the associated political tensions over resources could prove almost impossible.

As many have said, the population growth rate can be slowed more rapidly than current projections now suggest. The key is improving the quality of life, especially in rural areas of developing countries so that people don't feel that they have to have large families because some of their children are sure to die young or because many children are needed to support elderly family members. Population growth can be slowed by basic health care, family planning, and education services, coupled with expanded employment, improved natural resource management, increased equity in land ownership, and fair prices for agricultural products. The record amply demonstrates that much. If we also open markets in industrialized countries to the products of developing countries, it may be possible to sustain economic growth, minimize the increase in global population, *and* reduce the risk of rapid climate change.

Getting Personal

Everybody talks about the weather, but nobody
does anything about it.

EDITORIAL,
Hartford Courant, 1897

We Americans think we are pretty good! We
want to build a house, we cut down some
trees. We want to build a fire, we dig a little
coal. But when we run out of all these things,
then we will find out just how good we really
are . . .

WILL ROGERS

Only a few years ago, the greenhouse effect seemed a rather faraway—
if not far out—concern, more of a cocktail hour curiosity than a real
issue. In December 1983, shortly after a new report on the global
warming made a fleeting appearance in the media, *Vanity Fair* magazine ran a two-page cartoon picturing Manhattan as a tropical resort
with monkeys dangling from telephone poles, natives sporting colorful
costumes, and sidewalks sprouting palm trees. Such a distant prospect
seemed good for a laugh.

Today, the greenhouse effect is still a front-page story, it has become
an item on the agendas of governments and business, and it has entered
our everyday vocabulary. In 1988, *Time* magazine, instead of naming a
man or a woman as its "Man of the Year," chose instead to honor our
planet. "Now, more than ever," wrote *Time's* essayist, "the world needs
leaders who can inspire their fellow citizens with a fiery sense of
mission, not a nationalistic or military campaign, but a universal crusade to save the planet. Unless mankind embraces that cause totally,
and without delay, it may have no alternatives to the bang of nuclear
holocaust or the whimper of slow extinction."

More than any other environmental problem, global warming is
the product of ordinary human activities. Industry will probably have

to be regulated, but the public at large will also have to adjust. Ultimately, the problem is going to be solved by a lot of individual decisions. The question is how to change the individuals making those decisions.

Part of the problem is coming out the tail pipe of your car, for instance. By 2010, if nothing is done to improve the fuel efficiency of the world's 350 million carbon-spewing cars, these emissions will nearly double—to about a billion tons of carbon—and greatly speed up the global warming. But, according to Worldwatch Institute, if by that year there are no more than 500 million cars on the roads, averaging 50 miles per gallon instead of 19 or 20, carbon emissions from auto exhaust could be halved.

But what will rouse the public to act? The problems of global warming and ozone depletion seem somehow so overwhelming and intractable that, as one *Boston Globe* reporter writes, "most people's eyes glaze over" when they are even mentioned. "They feel they can do something to protect their drinking water, reduce radon in their homes and other indoor emissions and maybe even clean a toxic dump." But when it comes to global problems, "they wait for the government to act."

That, say environmentalists, is foolhardy. Americans can do much to reduce the use of fossil fuels and chlorofluorocarbons on their own. Even without any new legislation, people can adopt many of the energy-saving tactics that caught on temporarily after the last oil crisis. They can become informed consumers, buying products that don't deplete the ozone layer or contribute to emissions of greenhouse gases and trying to reduce pollution from their cars and homes. As voters, individuals can elect officials who take global warming seriously and put pressure on those in office to impose stronger efficiency standards. According to Rep. Claudine Schneider (R.–R.I.), author of one of several bills aimed at reducing global warming, "It will be action at the grass roots that moves Congress."

It was pressure from the public that pushed world leaders to phase out ozone-depleting chemicals by 1999, argues Richard Benedick, chief U.S. negotiator at the international meeting that approved the Montreal Protocol. "Even before the regulations on aerosols came into effect, the market dropped by half," says Benedick. "Never underestimate the power of the American consumer."

Although climate change has become a major global concern in a remarkably short time, a heated debate alone won't take us very far. Slowing the greenhouse warming and keeping Earth habitable will have to become everyone's problem. Indeed, the need for widespread involvement brings to mind the phrase of 1960s political activists, "if you're not part of the solution, you're part of the problem."

Fortunately, eliminating the causes of climate change—or at least tempering them—doesn't mean bringing technological progress to an end; it just means changing it. The "smart" technologies described in the last chapter—advanced, highly efficient, low-waste devices and techniques—will have to be substituted for the wasteful and polluting ones in wide use now. Forests will have to be preserved, and major reforestation efforts undertaken. And habits will have to change.

COMING AROUND

Recent opinion polls show that increasing numbers of people are concerned about the environment and that some are even willing to pay for it. In a *New York Times*/CBS poll taken in July 1988, some 65 percent of Americans surveyed agreed that "protecting the environment is so important that requirements and standards cannot be too high, and continuing environmental improvements must be made regardless of cost." A year later, pollster Louis Harris surveyed public attitudes toward the environment in 14 countries and found that people everywhere want governments to spend more to do more. Poll respondents said they would settle for lower standards of living and higher taxes if that's what it will take. Citizens of developing countries turned out to be even more concerned about the environment than those of industrialized countries, but—according to an earlier Harris poll—fully 81 percent of those polled in the United States would pay higher taxes for a cleaner environment.

At the same time, though, when quizzed on the greenhouse effect, many Americans seem bewildered. Despite the heavy media attention paid to the issue since 1987 or so, half of those polled by *Parents* magazine in 1988 knew little or nothing about it: 40 percent "said they had not heard or read anything about it," and only 50 percent of those questioned could actually identify what it was. "People tend to turn off when presented with technical information," suggests Michael Rampino of the Earth Systems Group at New York University.

Those polled didn't fear the greenhouse effect as an imminent threat. But once they grasped the problem, 88 percent said that we should start now to adopt policies that might enable humanity to cope with it. Even more encouraging, a majority said they were willing to change their habits, with 69 percent willing to drive less; 67 percent to take fewer showers, keep the house ten degrees colder in winter and ten degrees warmer in summer; and 61 percent to cut down on air travel.

These pledges give hope that people are beginning to see how their own interests (and their children's) are aligned with the planet's. There have been many instances throughout history of willing self-sacrifice for the sake of a higher cause. World War II is one of the best examples in our own time. Ronald Jager's *Eighty Acres* is an inspiring statement of what wartime commitments to austerity can do. Jager describes how people became accustomed to rationing as a fact of life:

> We came to assume that much of what we wanted was scarce or unavailable. Gas for the car could be had in small quantities; tires and tubes not at all. Near war's end Dad laced up a long tear in a tire's sidewall with bailing wire and buckskin to keep the tube from oozing out. It was a neat job but it didn't last, so he inserted a lumpy slab of rubber from a discarded tire. That worked better, but you could feel the thump of that lump in the tire like a bump in the road. Yet, it may have been as easy for rural people as for anyone to accept restrictions: The countryside was still feeling the Depression, was used to austerity.

Although they could significantly delay its worst effects, even austerity measures won't stop global warming at this point. But failing to curb our fossil fuels habit now will lock us into requiring "draconian cutbacks," as the Environmental Protection Agency has described them, in the future.

Leaving aside the difference between what people say they will do and what they actually do, individuals can't go it alone. It will take government incentives and regulations to make the public reduce fossil fuel use.

Even those who want an ecologically sound life style find that it's hard to go shopping with your conscience. If you wanted to stop using products with chlorofluorocarbons (CFCs), for example, you might not have any alternatives. Buying a new refrigerator for your home? You can't do it and avoid CFCs. But you can decide to buy a car without an air conditioner, improving fuel economy by about 2.5 percent—a double dividend. If you must have air conditioning for your home, you can buy a gas-filled unit that uses no CFCs. If you already have an air conditioner, you have a choice about how it is serviced and, ultimately, disposed of. And you have alternatives to buying a halon-based portable fire extinguisher.

Part of the problem is that our society is becoming so out of tune with nature that there are limits on what individuals can do. Our cars are built with styling and speed more than with energy efficiency in mind, and many U.S. cities are difficult to traverse on a bicycle. Many of our household goods are built to last only a few years or to "date" quickly. Since the term "throwaway society" was coined, such major cities as New York and Philadelphia have filled up almost all available landfill space with garbage. Our culture's taste for the ephemeral was perhaps best exemplified with the throwaway car.

In a market-driven economy like ours, energy-efficient products and services aren't likely to spring up unless there is a demand, and demand is conditioned by new regulations and programs, as well as energy prices and popular sentiment. It's the old chicken-and-egg problem. Unless governments start sending different signals, people will have few incentives to stop wasting energy and materials and to stop polluting.

The ultimate solution will come only with a major shift away from fossil fuels, which will take a government effort of unprecedented proportions and industry's cooperation. In the meantime, though, consumers and environmentally conscious people can do a lot to show their concern. What follows is a guide to the choices we do have now.

Living an ecologically sound life in the 1990s is a daunting task. It takes time, effort, and forbearance to find safe and healthy food, shop for energy-efficient appliances, separate your garbage, and, worst of all, pass up tempting conveniences. Then you find, for instance, that the best fossil fuels-saving home insulation may help cut the greenhouse effect down to size but also contain CFCs that gobble up the ozone layer. Twenty years ago people worried about phosphate detergents and a small list of pesticides. How much more complex the issues are now and how much more savvy consumers must be!

But it's not a lost cause, as Penny Ward Moser found. In an essay in *Life* magazine entitled "How I Saved the Earth in One Week," Moser describes her "brand-new, all-out effort to tread more lightly on the earth." "Last summer," she says, "I watched the crops bake to death on my family's Illinois farm. I saw garbage wash up on Long Island beaches. I saw the sad bald head of North Carolina's Mount Mitchell, whose trees were lost to acid rain. . . . Then one morning, after seeing the corpses of 300,000 fish that had suffocated when heat and drought took their midwestern lake, I became a born-again environmentalist. I missed the movement the first time around. Now it's back and I'm with it."

Moser starts out with a fair challenge—a house that "looks like Circuit City. . . . Two refrigerators. Enough high-tech lighting to shoot a movie. A kitchen with a computerized dishwasher so smart it argues with me about what cycle to use when. Three air conditioners, a 21-foot-long car and a Jacuzzi." But when she's done, she has learned to ride the bus; "eat lower on the food chain"; save every kilowatt, each one of which makes her power company "burn about 12 ounces—one Cracker Jack boxful—of coal"; and lower her water heater from 145 to 115 degrees. At the end of the week, Moser muses upon her new life style: "I don't mind hanging my wash. I'm getting faster with the flyswatter. I'm learning to like fluorescent lights." Phosphate-free detergent may mean a few spots on the dishes, "but if not using phosphates will help the oysters survive in the Chesapeake Bay, I'll cope with a few spots. I have to keep thinking this way."

Of course, the zeal of overnight convert Penny Moser might seem laudable to some, but her sacrifices would probably seem laughable to her counterparts in the Third World (and some here as well). Leaving aside Moser's Jacuzzi, most Third World women would consider themselves lucky to have such goods as clotheslines and dish detergents—and would find a lift in a bus a downright luxury.

In stark contrast, for example, consider the lot of Kenyan mother Susan Munoru, a Kikuyu, Kenya's most Westernized tribe, described in *National Geographic* as following "an endless round of long, burdensome treks." To get provisions for her family of six, she walks 2½ miles

Drawing by Ross © 1987, The New Yorker Magazine, Inc.

*"This isn't going to do the old ozone
layer any good, that's for sure."*

to the closest market town, returning by bus when she can afford it. For
fuel she uses charcoal sparingly, because it is hauled on foot from more
than half a mile away.

In the People's Republic of China, where energy consumption averages one-fifteenth that of the United States, families outside some major
cities use no electricity during the day because the province's growing
number of rural industries are higher on the list of priorities.

About one-third of the energy consumed in the United States is used
by individuals in their homes and cars. According to the editors of U.S.
News and World Report's *How to Cut Your Energy Costs*, "This is
energy consumption over which each of us has some control. Saving
energy, therefore, is something that lies in our power to achieve."

Since the late 1970s when President Carter urged Americans to begin
waging "the moral equivalent of war" and conserving energy to increase
our self-reliance, there have been some gains in residential energy
conservation. Indeed, residential use of fossil fuels is down almost 20
percent since 1976. During the past 15 years, U.S. efficiency in-

WHAT YOU CAN DO

- **Cut down on the use of cars and trucks** whenever possible and drive carefully to get better gas mileage. A car that gets only 18 miles per gallon—just 1 mile less than the current average for cars on U.S. roads—is estimated to produce about 50 tons of carbon dioxide during its lifetime. A car getting 27.5 miles per gallon—the new standard—will emit about 15 tons less.
- **Take the trains and use mass transit** where they are available and lobby for them where they aren't. One track of transit now carries as many passengers as six lanes of highway; if ridership grew, that same track could carry as many people as 15 lanes of highway, according to the National Association of Railroad Passengers.
- **Save on home heating and lighting.** Install better insulation and more efficient appliances; buy compact fluorescent bulbs; and urge power companies to make efficiency improvements and apply new planning techniques that keep the overall costs of supplying energy down.
- **Plant trees** and organize recycling programs in your community. Trees help cool cities, shade homes, and absorb carbon dioxide. Recycling saves on materials and fuel.

These are just a few of the steps individuals can take to save the planet and feel more in harmony with nature and with themselves. If the last generation gave us processed food, computers, and the ozone hole, perhaps the next generation can give us a better way of using Earth's resources.

---

vestments have cut the energy needed to produce a dollar of gross national product by 26 percent. But *total* energy use is up 8 percent since 1973, the upswing has been particularly great in the last 2 years, and a staggering amount of waste still exists.

If we have tempered home energy use, transportation is another story. According to studies cited by Environmental Action's Energy Conservation Coalition, oil consumption in U.S. transportation uses 1 million more barrels of oil per day than in 1973—that's not surprising considering that most vehicles on the road and in the sky can't use any fuel but oil.

Is it possible to gauge how much each person contributes to the greenhouse effect and ozone depletion? The world's five billion people put 5.6 billion tons of carbon into the atmosphere annually from fossil-fuel burning—more than 1 ton of carbon per person each year, on average. Americans put out about 350,000 tons of CFCs in 1986, or about 2.5 pounds per person.

This country accounts for a little over one-fifth of total carbon dioxide emissions, so each American's share is relatively greater—over 5 tons of carbon per person each year. In fact, the United States uses twice as much energy per capita as Europeans and the rest of the industrialized world to maintain a similar standard of living. As a *Washington Post* editorial noted in 1989, West Germans have smaller homes than Americans, but they live just as long and eat better.

GETTING THERE WITHOUT GUZZLING

"Drive a Chevrolet through the U.S.A.—the greatest country in the world!" So warbled a young Dinah Shore in a 1950s commercial. Although its record is improving and the world is catching up, "the U.S.A." is still leading the world in guzzling gas. Our mobile culture still drives highly inefficient cars and trucks worldwide and drives more of them (180 million total)—and farther—than any other industrialized country in the world. Whereas 135,431,000 cars were registered in the United States in 1986, there were 1,414,000 in India, 761,000 in mainland China, and 125,000 in Kenya, according to *National Geographic.*

For most of us, driving a car (or two) is one of our worst crimes against the planet. Auto sales have grown so much worldwide that cars have become a leading contributor to global warming as well as to air pollution. Worldwide, passenger cars consume one out of every six barrels of oil, according to the Institute for Energy Analysis. Even though the leading source of carbon dioxide is emissions from power plants burning oil, coal, and natural gas, automobile transportation is the biggest contributor to atmospheric pollution.

Transportation, overall—including buses, trains, and all other modes of transport—is also now "the largest and most rapidly growing drain on the world's oil reserves," according to *The State of the World 1988* by Worldwatch Institute. Nearly two-thirds of the oil that the United States

---

TOMORROW'S CARS TODAY

To slow global warming, we need cars powered by solar electricity, hydrogen, or other renewable fuels. Such vehicles would emit only negligible quantities of carbon and nitrogen oxides, and reducing the emissions of these pollutants from automobiles would be a boon to forests, human lungs, and such pollution-plagued cities as Los Angeles.

Sleek, shiny cars like the "sunracer" that General Motors raced in the international solar car race in Australia in 1987 are on many automakers' drawing boards. These cars would eventually evolve to accommodate roof-top solar cells, which would make these vehicles more practical.

Some carmakers are bringing out automobiles that go much farther than ordinary models on a gallon of gas. So buyers of new cars can go a long way toward reducing carbon dioxide by choosing these highly efficient vehicles.

Among cars now on the road, a number stand out for their impressive gas mileage. These cars use less gas because they are built with lightweight materials, which allow a smaller engine to be used without sacrificing performance; they use aerodynamic styling, which cuts wind resistance; they employ better transmission and engine designs, which convert less fuel into more power; and they use tires with less rolling resistance. The list includes:

Car	MPG	Size
Geo Metro	58	Subcompact
Honda Civic CRX HF	56	Subcompact
Ford Escort	42	Compact Car
Chevrolet Corsica	34	Midsize
Chrysler Lebaron GTS	34	Midsize
Dodge Aries	34	Midsize
Dodge Lancer	34	Midsize
Plymouth Acclaim	34	Midsize
Eagle Premier	31	Large

Cars in the prototype stage perform almost twice as well as these and get two to three times their gas mileage. Some examples

(continued)

TOMORROW'S CARS TODAY   *(continued)*

are the Volkswagen Auto 2000 (71 MPG), the VW-E80 (99 MPG), the Volvo LCP 2000 (81 MPG), the Renault EVE+ (81 MPG), and the General Motors TPC (74 MPG). Carmakers have perfected these prototypes and are developing many similar models. But they have held back from mass-producing them, they say, for fear that consumers won't buy them. According to Deborah Lynn Bleviss, author of *The New Oil Crisis and Fuel Economy Technologies,* many companies, such as Peugeot, are holding their prototypes in reserve until the next oil crisis. Consumers can make a difference by showing car companies that fuel economy is important to them, and by letting their legislators know that fuel economy standards should be strengthened.

---

uses is in transportation—more than it produces. "Worse," notes Worldwatch, "as auto sales grow, much of the world appears set on replicating this condition."

More and more people in the Third World are buying cars, with statistics showing that their fleets have grown twice as fast as those in the developed world, according to Worldwatch analyst Michael Renner. Their governments, he writes, have encouraged the trend, seeing the car as indispensable "as an engine of economic growth and a cornerstone of industrial development."

But while more and more people in Third World countries are becoming car owners, Renner cites studies showing that Americans use their cars far more. Quoting Australian researchers Peter Newman and Jeffrey Kenworthy, who surveyed 32 cities in Asia, Australia, North America, and Western Europe, Renner says that "on average, people in the highly car-oriented American cities use twice as much gasoline per capita as in Australian cities, four times as much as in European cities and ten times as much as in Asian cities. . . . U.S. cars travel some 1,259 billion miles annually—almost the same distance as all other cars worldwide taken together." Since that survey was done, Americans have logged many more vehicle miles: The number was 924 billion in 1970, but with the expansion of the nation's fleet, it soared to 1,128 billion in 1980 and to nearly 1,400 billion in 1988, according to Federal Highway Administration statistics.

Driving all these miles requires billions of gallons of gasoline. But that's not the only reason gasoline use has nearly doubled in the United States in the last three decades. The Energy Conservation Coalition points to a number of other factors behind these billions upon billions of gallons:

- Sales of light trucks, which are less efficient than automobiles, have grown by 30 percent since 1984, offsetting relatively flat car sales.
- U.S. gasoline consumption hasn't let up for 20 years, despite oil shocks and the fact that the country's oil reserves are dwindling, increasing pressure to find oil in hard-to-reach and environmentally sensitive places like Prudhoe Bay. Several observers point out that, in the case of the disastrous oil spill by the *Exxon Valdez*, Americans ought to be pointing the finger at their own wasteful habits as much as at corporate negligence. One environmentalist puts it like this: "If you don't drive a car that gets at least 35 miles to the gallon, don't tell me how outraged you are about dead otters."
- More people are driving in congested cities, where frequent starts and stops translate into low fuel economy.
- Standards for fuel economy are low.
- The real price of gasoline is below what it was in 1973.

Since 1973, the average efficiency of *new* cars has doubled. According to the Energy Conservation Coalition, "a major reason for this increase is the Corporate Average Fuel Economy (CAFE) standard passed in 1975." Nevertheless, with more cars and light trucks on the road, gas consumption for automobiles and light trucks has gone from 6.1 million barrels per day in 1975 to 6.8 million per day in 1987.

Many fuel-efficient models have reached the test tracks in Europe and Japan. Some are already on the road, like the Honda CRX, with its plastic body panels and aluminum parts, and Subaru's little Justy, with its innovative transmission. By contrast, U.S. carmakers are, as World Resources Institute's Jessica T. Mathews puts it, "allergic to innovation." And though the regulatory climate began changing again in 1989, with prices so low, fuel economy has been a relatively low priority for U.S. consumers.

CARS AND CHLOROFLUOROCARBONS

Car air conditioners add to the load of chlorofluorocarbons when they leak. These "mobile air conditioning" systems, as they're called, are the single biggest source of CFCs in the upper atmosphere.

WHAT YOU CAN DO

- **Drive an efficient vehicle.**
- If you're buying a new car, buy one that gets at least **35 miles per gallon.** "There's no reason to get a car that gets so few miles per gallon when you can get a car that you can drive much more efficiently," says Brady Bancroft of Rocky Mountain Institute. "Look for the 30s at least."
- Van pool, car pool, take public transportation if you can.
- Keep your car as well-tuned as possible.
- If you have an air conditioner, make sure you **recycle the CFCs** with the "vampire" machines. Try to get your local gas stations to do CFC recycling, and urge auto supply shops to stop selling the 14-ounce "refill-it-yourself" cans of CFCs.
- Help environmental advocates **lobby for high standards** for fuel efficiency and emissions. Join environmental organizations and campaigns, write your members of Congress, and get local publicity for the issue.
- **Walk or bike** where you are going whenever you can. Neither contributes to the greenhouse effect, and both contribute to fitness.
- **Combine errands** into one trip.
- **Don't idle** your engine. Turn it off rather than let it run for more than one minute.
- **Keep tires inflated** to the proper pressure level.
- Don't make jackrabbit starts, which waste fuel.
- **Drive more slowly.** Traveling at 65 MPH instead of 55 MPH will lower your gas mileage by more than 15 percent.
- Don't carry unnecessary weight.
- **Don't rev** the engine before it is shut off.
- Make sure your car's front wheels are in alignment.
- Resist the temptation to turn on your car's air conditioner, which lowers efficiency 2.5 percent.

Carmakers will be switching from the most ozone-depleting CFC (CFC-12) to less harmful substitutes during the next 10 years. Meanwhile, producers plan to begin a voluntary CFC-recycling program to make sure that the CFCs are conserved—and contained.

Until industry voluntarily began developing standards for recycling CFCs in January 1989, auto mechanics and their customers had routinely vented the insulating chemical into the atmosphere—no mere *pffft*, considering that an estimated 11,000 U.S. businesses currently service car air conditioners. If more gas service stations bought recycling equipment, fewer CFC leaks would occur. "Ask local gasoline service stations to buy recycling equipment to recycle CFCs from car air conditioners," recommends Liz Cook of Friends of the Earth, an environmental group with affiliates in many countries.

To limit emissions from mobile air-conditioning systems further, Friends of the Earth is seeking a ban on the portable "do-it-yourself" cans of CFCs sold to drivers for refilling their car air conditioners. These cans are prone to leaks and can easily be mishandled.

HOME ENERGY USE

Our homes consume most of the electricity we use. Yet most of the heat generated is wasted. Amory Lovins of Rocky Mountain Institute notes that most American houses have 2 square yards of leaks and cracks. Between 40 and 70 percent of our heating expenditures are wasted on heating and cooling outside air that unnecessarily mixes with the inside air. Obviously, home is where people can eke out the most savings with conservation measures and common-sense thrift.

After the energy crisis of 1979, countless reports from scientists and engineers encouraged Americans to plug up the leaks in their homes. State energy offices around the country issued guides to consumers about how to weatherize their homes, how to install storm windows, weather stripping, solar panels, etc. Many states offered free energy audits to customers using particular utilities—and still offer them.

Hundreds of books were written on energy conservation in the late 1970s and early 1980s, from books on window caulking and weather stripping to energy-saving cookbooks. But just as conservation picked up momentum across the country a decade ago, nearly all the energy-conservation programs launched in the late 1970s were dismantled.

Nevertheless, every state still has an energy office, and a national service offers consumer information and tips: the Conservation and Renewable Energy Inquiry and Referral Service, which has a toll-free number (800-523-2929). By law, all utilities must offer energy audits for $15 per household.

Best estimates are that Americans can still achieve significant savings, of money and energy, through simple conservation measures at home. The number of people retrofitting their homes has fallen with energy prices, but officials at the Conservation and Renewable Energy Inquiry and Referral Service have seen an upsurge in the past few years in the number of people calling for information. "There's a steady interest in energy conservation and renewables, from the responses we get here," says Paul Coombs, response analyst. His personal observation is that people remember the energy crises of the past and know that the cheap energy prices today could shift in the 1990s.

How much you can cut your energy bill as a homeowner or apartment dweller depends on how much you now pay for fuel and how many improvements you make. Costs vary from state to state and region to region, but the rule of thumb is that electricity is the costliest energy form you can buy, oil the next most costly, and natural gas the cheapest.

Saving on Home Heating

Air enters the house—and heat escapes it—through cracks around windows, door jambs, fireplace dampers, and other openings in walls and floors. How much heat is lost, and where, varies from house to house. One study done in Minnesota indicated that 27 percent of the heat lost as cooler air infiltrates from outside goes out doors and windows; walls lose 15 percent; roofs lose 13 percent; and 5 percent escapes through the floor slab under the house.

Besides plugging holes, most experts agree, turning to simple energy conservation measures such as insulating your water heater and hot water pipes and maintaining your furnace can pay for themselves in saved dollars in a fairly short time.

Hot Water

A second way that you can make a difference is with the hot water heater, the next biggest electricity user under your roof.

WHAT YOU CAN DO

- **Caulk and weatherstrip.** Sealing up the leaks in windows, doors, and other openings is perhaps the first and simplest measure that can be taken. Energy-saving guides suggest testing windows and doors for airtightness. "Move a lighted candle around the frames and sashes of your windows," writes Jean E. Laird in *The Homemaker's Book of Energy Savings*. "If the flame dances around, you need caulking and/or weatherstripping." Other places to check for leaks: where windows are joined to door frames, where porches meet foundations, and where the basement meets the foundation.
- **Install better windows.** Because a single pane of conventional glass offers little protection from the cold, experts suggest using double-glazed or triple-glazed windows or buying storm windows (or combination storm/screen units). However, new single-pane windows with "low-emissivity coatings" are just as effective insulators as existing double-pane windows.
- **Service the furnace.** Ask your service person to check the efficiency of your furnace, if it's not already done as part of your service contract. It's more efficient to burn oil in your oil burner than to buy electricity from a coal plant for an electric space heater.
- **Check your fireplace.** Its romantic light can be expensive in energy terms. Make sure it's efficient because how you maintain it and whether you add glass doors or a convection grate can make a difference of a factor of eight in how much wood you have to use.

  To get the most out of your fireplace: close the damper when it's not in use; have your chimney cleaned at least once a year, and more often if soot or creosote has built up; consider putting on glass doors, heat circulating devices, and other products now on the market that will improve your fireplace's operating efficiency.

WHAT YOU CAN DO

- **Install an insulator around your water heater,** either a blanket or jacket. They cost about $20 and are available at stores that sell hardware, plumbing supplies, or building materials.
- **Install low-flow shower heads,** which use only half as much water as their conventional counterparts, and low-flow sink aerators in all the faucets in the house. Because you will use less hot water, your water heating bill will fall dramatically. Shower heads typically cost about $10 each, aerators about $1.
- **Turn the temperature down** when you won't be needing your water heater, which keeps the water hot 24 hours a day. Why heat water when you're away for the weekend or on vacation?

Lighting

The third major way you can save energy is in lighting, which consumes more than 20 percent of all electricity used in the home, according to the Department of Energy. How much lighting is enough? Experts advise allowing 1 watt of incandescent light per square foot of floor space, or up to 3 watts per square foot if your ceilings are very high and the walls are dark.

The incandescent light was the original light bulb. But fluorescent lighting is three times more energy efficient. One 40-watt fluorescent lamp gives off more light than three 60-watt incandescent bulbs, and now there are "compact fluorescent bulbs" on the market that use 75 percent less energy than the ordinary light bulb. And some new-fangled fluorescents don't give off the bluish light that some people associate with schools, hospitals, offices, and other institutions. Replacing a single 75-watt incandescent bulb with an 18-watt fluorescent bulb can save about 385 pounds of coal over the lifetime of the bulb. High-pressure sodium, low-pressure sodium, and metal halide bulbs are a few of the alternatives to standard bulbs.

Energy-efficient light bulbs are more expensive than conventional

WHAT YOU CAN DO

- **Keep all lamps and fixtures clean.**
- **Reduce overall lighting** in nonworking areas by removing one bulb out of three in multiple-light fixtures and, for safety's sake, replacing it with a burned-out bulb.
- **Use one large bulb instead of several** small ones where bright light is needed.

bulbs, but they will last eight to fifteen times longer and cost far less in the long run, according to the American Council for an Energy-Efficient Economy. And you don't have to make the switch all at once. "Find the two to four lights in your house that you use the most. Maybe the one in the kitchen, or that big one in the living room," says Brady Bancroft. According to Bancroft, the most efficient light bulbs are widely used by businesses, but they haven't yet caught fire with residents.

In the Kitchen

After space heating and cooling—the number one use—and heating water, preserving food is the third largest consumer of energy in the house. Refrigerators operate non-stop, so they use a good deal of energy. Old refrigerators can be particularly energy hungry.

Some of the newer refrigerators are much more energy efficient, by a large margin. Some new models use only 200 kilowatt-hours per year. Others use up to 700 kilowatt-hours, compared to 1,400 to 1,600 for traditional models.

Select a new refrigerator with the same kind of care you devote to buying a new car. The larger the model, the more energy it will require. And if it is too large, and you can't keep it filled, it will take even more energy to keep cooled. Likewise, stay away from refrigerators with extras like ice-cube makers, cold drink dispensers, and other frills. According to the authors of *How to Cut Your Energy Costs*, "A refrigerator can be a very simple appliance costing as little as $25 a year to operate. However, when it is made into a convenience center . . . it can become the third or fourth largest energy consumer in your house—especially if it is 30 to 50 percent oversized, as many units now are."

WHAT YOU CAN DO

• **When buying a new refrigerator, buy one with a power-saver switch.** It will turn off unnecessary power, like that for the heating element that prevents the refrigerator from sweating on the outside—a device that doesn't need to be working all the time.

• **Place your refrigerator in a cool location** for maximum efficiency. Next to an uninsulated wall or in the sun, the machine will only have to work that much harder, especially in summer.

While we're on the subject, Amory Lovins has plenty to say about the American refrigerator circa 1980. He looks at the anatomy of an appliance that got bigger and badder. "Around the end of World War II," writes Lovins in *Life after '80: Environmental Choices We Can Live With*, "your refrigerator motor was probably 80 or 90 odd percent efficient and it sat on top. Nowadays the motor is maybe 50 or 60 percent efficient, probably because the price of electricity to your house has dropped severalfold since then; and the motor sits underneath, so the heat goes up into the box. Therefore, your refrigerator probably spends about half of its effort taking away the heat of its own motor."

The next step, Lovins says, was that manufacturers began to skimp on insulation. "It got thinner and thinner because they tried to make the inside pretty big compared to the outside. . . . Because of that and because it is designed so that when you open the door all the cold air falls out, it frosts up, so your refrigerator probably has in it a lot of electric space heaters that go on now and then to defrost it. And then it probably has electric heaters around the door to keep the gasket from sticking because manufacturers cannot be bothered to use a Teflon coating. Then the heat gets pumped out the back into a kind of radiator, which is usually pressed right into that thin insulation to help the heat get back inside as fast as possible. In addition, the refrigerator is probably installed next to your stove or dishwasher, so when that goes on, it goes on. It is really hard to think of a better way to waste energy."

APARTMENT LIVING

If you live in an apartment rather than a house, you probably want to be spared talk about wrapping your hot water heater and installing passive solar window panels. You don't own the place, so why should you care? As environmental activist Peter Harnik writes in *Environmental Action* magazine, "Owning a home may be the American dream, but for those of us who are still dreaming it instead of living it, all those glowing articles about 'the ecological house' and 'the self-reliant homestead' can be grating."

But you don't need to own a house to practice ecological living. You can do it when you rent. As Harnik writes, apartment dwellers "need a household ecology ethic of their own."

There's a sound economic reason for adopting such an ethic. In most cases, tenants—not landlords—pay the gas and electric bills (and sometimes even the water bills), so renters obviously stand to gain from conservation improvements. Renters are also the major line of defense against energy inefficiency. Apartments tend to be more energy wasteful than homes for several reasons. For one, "rental units are more likely to be in poorer structural shape than owner housing, and less likely to have many of the conservation features that were built into structures in the post-Embargo era," write Deborah Lynn Bleviss and Alisa Ayn Gravitz in *Energy Conservation and Existing Rental Housing*. Apartments also "tend to have a lot of windows on the wall facing out (or at least you hope so), and that outer wall usually has poor thermal performance," notes William Prindle of the Alliance to Save Energy.

Renters account for a third of all residential energy use, and, Bleviss and Gravitz write, "up to one-half of the energy consumed by rental structures could be saved through cost-effective conservation techniques." Not surprisingly, twice as many homeowners as renters have sought to maximize energy savings by installing insulation, storm windows, and other energy-efficient features, according to the Department of Energy. From 1978 to 1984, average household energy use dropped by 15 percent, DOE figures show, but the savings went primarily to higher-income households.

In a 1988 study by the Alliance to Save Energy analyzing these trends, *Making Housing More Affordable through Energy Efficiency*, Prindle and Michael Reid write that multifamily buildings stand to gain anywhere from 12 to 42 percent in energy savings through retrofitting and that it could, especially for lower income families, "make the difference between affordable and unaffordable housing."

More than a quarter of the country (27 percent), or 28 million households, live in apartments, and energy is a major expense for them. Low-income tenants, according to the Department of Housing and Urban Development, pay up to one-third of their incomes for energy bills. In some cases, the poor may spend more on electricity than on rent.

Yet, even though tenants stand to gain by reducing their energy bills, they have little incentive to invest in major improvements to make their apartments more energy efficient. At the same time, landlords have little incentive to retrofit apartments at the tenants' request because they have no assurance the tenants won't move out. "Almost half of renters live in their apartments less than a year; more than three-quarters occupy their units for less than five years," write Prindle and Reid. "With such short time horizons, tenants are not interested in funding (through their rent payment) improvements that may not pay back for several years. By the same token, owners are unlikely to make substantial energy investments on behalf of tenants who can be expected to leave within a few years."

As a solution to this double bind and other inherent obstacles to conservation in rental housing—the adversarial relationship between landlord and tenant, barriers to financing, the fact that owners can deduct energy bills but must depreciate energy improvements over a long period—energy and housing activists are trying to bring in a third party, such as the local utility or a community action program, that can help encourage conservation. "They put pressure on all the parties to put commitment into energy conservation," says Nancy Hirsh of Environmental Action Foundation's Energy Conservation Coalition.

As mentioned in chapter 4, some utilities are now sponsoring a number of programs, which vary by state and utility service territory. They include appliance rebate programs (in which utilities pay the difference between a higher-priced, efficient appliance and a cheaper, less-efficient one) and financing help for such conservation investments as wall and window insulation.

Pacific Gas and Electric Company in California, for example, will replace windows, put in insulation, and make envelope improvements costing up to $3,500, and they will finance loans. On the East Coast, Northeast Utilities, in partnership with the Conservation Law Foundation of New England, is seeking out homes and rental units where conservation gains could be most dramatic. According to Armone Cohen, an attorney with the Conservation Law Foundation, Northeast

WHAT YOU CAN DO

- **Write your landlord.** "Convey to your building owner your sentiments about the need to conserve our limited energy resources," writes New York State's Cooperative Extension Service. "In person or by letter, point out how energy is being wasted in your building."
- **Keep your thermostat** down during the winter heating season, between 65 and 68 degrees during the day and 60 degrees at night.
- **Caulk around your windows and doors** to reduce heat losses. Plasticized roller shades, tightly sealed at the edges, can help too. Use insulated draperies, curtains, or interior shutters. Keep furniture away from radiators.
- Install your air conditioner in the window most protected from direct sunlight. If you want your room cooled when you get home, connect the unit to a timer set a half-hour before you return. In summer, set the thermostat at 78 degrees or above.
- To save on water heating, **take quick showers** instead of baths, run only full loads of laundry, and run a full dishwasher.

Utilities is covering the costs of energy conservation measures up to 100 percent, and an average grant to homeowners is about $1,200 to $1,500.

"Utilities started by sponsoring public information programs and audits," says Hirsh, "but in the last three years or so, they are really spending dollars." Their commitment to conservation is increasing, she says, because regulators are denying utilities' requests to incur high construction costs. As a result of this movement, there are now financial incentives for consumers—apartment dwellers as well as homeowners—to invest in energy conservation.

Tenants should ask their local utility what programs are available or contact their state energy offices. With this knowledge in hand, they should then talk the building owner into making as many energy-conservation investments as possible. With the help of utility financing programs, some of these measures—storm windows and storm doors,

---

FOR MORE INFORMATION

Tips for an Energy Efficient Apartment (Stock #061-000-00658-3), U.S. Department of Energy, Office of Conservation and Renewable Energy, Building Energy Research and Development, Building Services Division. For sale by the Superintendent of Documents, U.S. Government Printing Office, Washington, DC 20402 ($1.00).

Making Housing More Affordable through Energy Efficiency, William Prindle and Michael Reid, The Alliance to Save Energy, 1925 K Street, N.W., Suite 206, Washington, DC 20006 ($5.00).

At Home with Energy, American Association of Retired Persons, Consumer Affairs Division, 1909 K Street, N.W., Washington, DC 20006 (free).

---

insulation, and separate meters and thermostats—may be less costly to the owner than expected.

Community action organizations continue to support federal weatherization assistance programs for low-income households enacted a decade ago, hiring the contractors to install insulation, weather stripping, etc., and making sure they complete the job.

But many low-cost and no-cost measures can be undertaken by tenants themselves. Some of the best are turning the thermostat down and installing shades.

There are some Catch-22s. Although foam insulation is very efficient, CFC-blown foam insulation helps deplete the ozone layer. You will probably have a hard time finding out just how the insulation for sale is made, but you can always lobby to get manufacturers to disclose it on the label.

LONG-TERM STRATEGIES FOR ENERGY CONSERVATION

Once you've cut your demand for electricity, then you can look to the supply side. Some of the best big-ticket investments are a solar water heater or a sun space on the south side of the house. If you live in a remote location, you could even hook into a photovoltaic generator. The price for photovoltaics (solar cells) is dropping faster than anyone could have imagined a few years ago, thanks to technical breakthroughs in the late 1980s. Brady Bancroft predicts, "In the 1990s, they should be a big source of energy."

DOING MORE WITH LESS

In the 1970s, a catchphrase of energy analysts and activists was "getting more bang for a buck." Today, this quest is even more important, and one key for the consumer is buying energy-efficient products that last. Despite the ubiquity of ecologically unseemly products—from polystyrene to disposable diapers—it is possible to shop wisely with the greenhouse problem, in particular, and environmental integrity more generally in mind. And although it seems as though there are ten energy-intensive products for every efficient one, appliance efficiency ratings can be a good guide. Take careful note of the energy-guide ratings that are required by law on many appliances.

Virtually every appliance made today uses some energy, but though some can be highly wasteful (such as most dishwashers, clothes dryers, refrigerators, and water heaters), others are highly efficient. Using a toaster to make toast or a waffle iron to make waffles makes better sense than heating up an oven or a stove to do the job. (As Amory Lovins puts it in another context, "Why cut butter with a chain saw?") And an electric shaver, by one estimate, takes a year to use as much energy as it takes to heat a week's supply of hot water for lather shaving. Ceiling fans, like the ones immortalized in the movie *Casablanca*, use no more energy than a 15-watt light bulb and can make a room feel cool even in sweltering temperatures. Energy saving may or may not be the reason they became all the rage in the 1980s, but the vogue means that they are readily available—often, at a discount.

In general, appliances are becoming much more efficient. This trend is due mainly to appliance-efficiency standards, says Liz Kellenbenz of the American Council for an Energy-Efficient Economy, which publishes the energy-efficiency ratings of most of the major U.S.-made appliances on the market. More and more appliances, says Kellenbenz, are being made with energy-saving features. Many dishwashers, for instance, now have "overnight dry" settings that automatically turn the machine off once the dishes are rinsed, saving as much as a third of the energy needed to complete a cycle.

Besides efficiency, a key issue in an era of finite resources is how long products will last. Investing care and money in a few possessions

WHAT'S YOUR E.Q. (ENVIRONMENT QUOTIENT)?

The United States puts out 1.3 billion tons of carbon annually. With a U.S. population of 250 million, that's 5.2 tons per person per year. Granted, much of this total comes from industry, but we can each reduce our portion of carbon output by making everyday decisions to use energy in a way that produces less carbon. By rating yourself on the following chart, see how many ways you *do* or *can* use energy to slow global warming! (*General rule of thumb:* If you save 1 kilowatt-hour, that represents a savings of 0.4 pounds of carbon not released to the atmosphere.)

	Energy Savings	Reduction in Carbon Emissions
Home		
Improving insulation in your hot water heater	300 kilowatt-hours per year (kwh/yr.)	120 lb.
Switching from resistance heater to heat pump	2,000 kwh/yr.	800 lb.
Switching from typical refrigerator/freezer to more efficient model	1,250 kwh/yr.	500 lb.
Updating central air conditioning	1,000 kwh/yr.	400 lb.
Substituting an 18-watt compact fluorescent light for a 75-watt regular bulb (8 hours/day)	170 kwh/yr.	70 lb.
Car		
Carpooling instead of driving alone, for five friends	1,000 gallons gasoline/yr.	5,000 lb.

(continued)

WHAT'S YOUR E.Q.? *(continued)*		
Taking inter-city train instead of air flight	2 gallons oil per 100 miles	10 lb. per 100 miles
Not driving (the average U.S. car goes 10,000 miles per year)	500 gallons	2,650 lb.
Driving a car with 30 m.p.g. (instead of car with 20 m.p.g. for 10,000 miles)	167 gallons	880 lbs.
Driving a car with 40 m.p.g.	250 gallons	1,320 lb.
Driving a car with 50 m.p.g.	300 gallons	1,580 lb.
Tuning up your car (at average 500 gallons/yr)	50 gallons/yr.	265 lb.
Community Using a push mower instead of power lawn mower	5 gallons gasoline/yr.	26 lb.
Planting trees to the south and west of your home	500–1,500 kwh/yr.	1,000 lb/yr. (from both carbon stored in trees and energy saved from reduced air conditioning)

Recycling	Cuts Pollution
Recycling glass	22%
Recycling paper	73%
Recycling aluminum	95%

that will endure "not only makes sense financially," writes Roger B. Yepsen, editor of *The Durability Factor*, "but also happens to be a first step in fashioning a contemplative or spiritual life." From Ivan Illich to Warren Johnson and Wendell Berry, other philosophers of daily life have made the same point: not all modern conveniences are really convenient if you add up the time it takes to earn the money to buy them and keep them running, to make space for them, and to replace them periodically.

But while durability seems to be an obvious good, it's harder and harder to find. There seems to be a fashion for goods with long-term value; running parallel with it is a rage for junk. "The sheer quantity of goods, of marvels and gimmickry, has been exhilarating, but quality hasn't gotten a boost," writes Yepsen. "Our basements, yard sales, and dumps are gorged with radios, refrigerators, cars that died young and still shine as if new."

A more important issue than clutter and junk is the trade-off between resisting the temptation to buy into instant obsolescence, on the one hand, and contributing to the market for new energy-efficient goods, on the other. Turnover leads to waste but also to technological innovation. Here again, consider the case of the car. Right now Americans keep new cars an average of 5.6 years—55 percent longer than they did just a decade ago, according to the Automotive Information Council. "Customary environmental thinking is to promote the durability of goods (cars, etc.) and in general that goal makes good sense," says World Resources Institute president Gus Speth. "But if the objective is to transform technology (for instance, to move to cars getting 100 miles per gallon over a period of a few decades), then this requires turnover in stocks (new or recycled cars)." Resolving this issue isn't easy, and consumers will have to make car-buying decisions on the basis of the best information they can find about specific models.

What all this has to do with the greenhouse effect is that the more short-lived objects there are, the more material and energy resources are depleted foolishly and the more useless objects—usually plastic—there are to clutter the planet. It's up to consumers to demand well-crafted, long-lived products. Manufacturers give us what we ask for, argues Yepsen, and "if people demanded no-nonsense products and would back up their words with dollars, then companies surely would produce them."

WHAT YOU CAN DO

* Urge your town or city to **phase out fast-food and plastic packing** throughout commerce, following the lead of Minneapolis and a number of other cities.
* **Avoid buying CFC-containing products,** including foam cushions and mattresses.

COMMUNITY AFFAIRS

Global warming is too big a problem to be solved by the isolated efforts of communities and neighborhoods, but close to home is a good place for individual efforts to start. It is here that volunteer and municipal groups do the vital work often neglected at the national level, like recycling and tree planting.

Tree Planting

Trees have always been appreciated for their greenery, shade, and beauty. But more and more in recent years, trees have come to be valued for their help in offsetting pollution, especially in cities. In the 1990s, they will be put into service to combat the greenhouse effect.

Trees cool overheated cities in the summertime by as much as 7 degrees Fahrenheit, shading buildings and cutting the need for power to run air conditioners in small buildings. They thus reduce the demand for fossil fuels. Most of all, they absorb carbon dioxide from the atmosphere, and give off life-supporting oxygen. A team of energy specialists at Lawrence Berkeley Laboratories studying urban forestry believes that planting trees is "the cheapest way we know of saving energy and money and reducing carbon dioxide."

Arbor Day comes each spring in different months throughout the country, but in 1989 the annual event was given an infusion of life with the American Forestry Association's (AFA) "Global Releaf" program, which called for us to plant 100 million trees in our communities by 1992 as a way to fight the global warming. Urging people to "plant a tree, cool the globe," the group kicked off activities in three dozen cities.

There are only so many times you can recite Joyce Kilmer's poem, says forestry activist Deborah Gangloff of AFA, who usually sends out a thousand Arbor Day kits each year to civic and municipal groups. In 1989, though, she says, the response to the global warming plight was overwhelming, and the group received thousands of calls and letters pledging to plant trees.

Planting trees is a way to get people involved in maintaining the environment, says Gangloff. So far, tree-planting fever has spurred several local crusades. A hand surgeon in Kent County, Michigan, has planted 10,000 trees, underwriting the $60,000 cost himself and challenging the county to match him with $90,000 more. Dr. Alfred Swanson, chief of orthopedics at Blodgett Hospital in East Grand Rapids, planted large (5- to 6-foot) trees with the help of local nursery staff and 300 fifth graders in the East Grand Rapids School District. Jeanne Swanton, a science teacher at Fairfield High School in Fairfield, Connecticut, helped her classes plant 68 trees on the school grounds. Students from an architectural drafting class drew the scale drawings of the area, and the artistically inclined drew posters and flyers. It was the first time she got her students "to get excited about anything in science," she told AFA.

There are 20 million acres of trees in what activists like to call the urban forest. They are street trees in cities and suburbs, park trees, and trees on large, private tracts. Increasing the stands of these trees will be doubly beneficial because these trees can actually change microclimates, and they can help slow the greenhouse effect. According to H. Akbari of Lawrence Berkeley Laboratories, trees can break up "heat islands" by shading buildings and concrete. Strategically planted trees can reduce interior energy use by 10 to 50 percent.

Trees clean the air, provide shade, and muffle noise. But tree planting certainly isn't a solution to global warming by itself, activists caution. To keep up with the amount of carbon dioxide going into the atmosphere, according to World Resources Institute, Earth would need at least 3 billion more acres of trees to keep atmospheric carbon dioxide in balance. That's enough forest to cover the entire 50 states with about 700 million acres left over.

Nevertheless, planting 100 million trees in the United States, foresters say, would offset carbon dioxide emissions by 18 million tons a year, at the same time saving consumers an estimated $4 billion a

WHAT YOU CAN DO

* Whether it's in your backyard or a forest in your city or town, **plant a tree.**
* Monitor the National Forest Service at home and the world development banks abroad to **work for global forest preservation.**
* **Encourage government** to enhance urban foresters' ability to manage tree preserves.

year in energy costs. (Maybe Kilmer's ode *can* be recited once more!)

Recycling

In the United States, 150 million tons of garbage are generated each year. Of that, the average U.S. household throws out 1,800 pieces of plastic; 13,000 pieces of paper; 500 aluminum cans; and 500 glass bottles each year. Much of this waste could be recycled, saving energy and materials (many of them toxic) and reducing the need for fossil fuels in raw material extraction, processing, and transport.

Using less energy will naturally reduce greenhouse gas emissions. But it will also cut other air pollution as well. According to a National Appropriate Technology Assistance Service study, using recycled instead of virgin materials would cut air pollution by up to 22 percent in the case of glass manufacturing, use, and disposal. For paper, the figure is 73 percent; for aluminum, 95 percent.

Recycling can help avoid another garbage-generated threat to global climate—incinerators with no energy-recovery technology. "Incinerators emit carbon dioxide just as any other combustion process does," writes Jeanne Wirka in *Environmental Action* magazine. "In fact, current incinerator technology aims to maximize carbon dioxide emissions, because this means a more efficient burn with less carbon monoxide—which is, unlike carbon dioxide, a regulated air pollutant." Under the banner of "not in my backyard," communities have battled the construction of mass incinerators because of the toxic chemicals they emit; global warming should give these communities further ammunition and a new battle cry—"not on my planet."

WHAT YOU CAN DO

- **Find out about your community's recycling program** and sort recyclable waste (newspapers, glass, aluminum, etc.). What's left? Use vegetable scraps for compost; then fertilize your garden.
- **Urge town or city officials to set up curbside pickup** if it isn't yet an option in your community. Pressure officials to launch a program that regularly picks up such household toxic materials as paints and solvents, furniture polishes, and oven cleaners. At a minimum, have them designate a drop-off point where conscientious consumers can bring them.

Environmentalists argue that starting a recycling program is far preferable to incineration. Nine states, including New York, New Jersey, and Connecticut, according to the Keep America Beautiful organization, have mandatory recycling laws. But not all of them have regular curbside garbage collection of recyclable materials. (From a greenhouse perspective, the choice of how to dispose of garbage isn't simple. Burning garbage releases carbon dioxide. So does recycling—but less. Landfilling garbage releases methane—a more potent greenhouse gas than carbon dioxide.)

Thanks to population growth and urbanization, human beings are also living much closer together on smaller amounts of land, squeezing out forests and wildlife. Although not yet as dense as in most other countries, the people-to-land ratio is getting tighter in the United States. With the continental U.S fixed land area of 2.27 billion acres and increasing population, writes Martin Pawley in *Building for Tomorrow*, the acreage per person is steadily falling. "In 1800, it was 450, in 1900 down to 35, in 1970 down again to 11, and by the year 2000 it will be only 7."

According to Pawley, the U.S. economy consumes about 6 billion tons of fuels and raw materials every year to produce 300 million tons of food, 250 million tons of major manufactured materials (paper, metals, glass, textiles, plastics, and rubber), and several trillion kilowatt-hours of electricity. Yet, says Pawley, "only about 1 percent of the raw materials fed into this vast productive mechanism is retrieved by the resource recovery industry."

The Global Greenhouse Network writes that local organizations can also sponsor recycling efforts: "Recycling drives are an excellent source of income for local organizations. Paper and aluminum can collections can support everything from the local high school band to Boy Scout troops and Little Leagues." Although markets for recycled paper are temporarily glutted, its very availability at low prices should mother innovation on additional uses for this resource, which, in turn, should eventually expand these same markets.

TENDING OUR GARDENS

Changing how we use our homes and drive our cars is the major contribution that individuals can make to reducing greenhouse gas emissions. But some of the smaller, mere symbolic steps are also worth considering. Just as some people are switching to smaller cars, keeping their thermostats turned down, and getting serious about other forms of energy conservation, some are taking the same approach in the garden. A growing number of people are reducing the size of their lawns, or even eliminating them, instead planting native shrubs and wildflowers that don't require much fertilizer or pesticide.

"Put your mower in mothballs," says William Niering, professor of botany at Connecticut College and director of the Connecticut Arboretum. Someone who once boasted he had "the loudest lawn mower" in his suburb, Niering has now turned his mower into a plow. Instead of a conventional suburban lawn, his half-acre has been transformed into a patchwork of gardens and naturalistic landscaping much like a meadow filled with wildflowers.

Americans use 3 million tons of fertilizer on 5 million acres of lawns each year, writes Niering in *The Connecticut College Alumni Magazine*. We also use some 40 million lawn mowers that consume more than 200 million gallons of gasoline each year—enough to contribute 1.9 million tons of carbon dioxide to the atmosphere annually. Niering's idea is that you should reduce the size of your lawn or even do away with it. Not that lawns aren't beautiful, he says. They are "aesthetic, artificial creations of man, but environmentally very expensive to maintain." The advantages, says Niering, are that naturalistic yards usually need little

WHAT YOU CAN DO

- **Use hand tools instead of power tools,** especially for small jobs. One kilowatt-hour of electricity saved will reduce carbon dioxide emissions at a coal-fired power plant by more than 2 pounds.
- **Switch to wildflowers and native plants.** The Connecticut College Arboretum has suggestions and information on varieties that can be planted on the East Coast. Its address is Connecticut College, New London, CT 06320. For information on the West Coast, contact Strybing Arboretum, Golden Gate Park, San Francisco, CA 94117.

maintenance because they are already well-adjusted to the regional environment, and they recycle their own nutrients by littering their leaves, thus cutting out the need for fertilizers or pesticides.

Niering is not a solo advocate for care-free landscaping. Take botanist Paul Mankiewicz, a consultant to the New York Botanical Garden and director of the Gaia Institute in New York City. He also argues that reducing the area of land given over to lawns and instead planting native plants and shrubs could help reduce global warming—the deeper root systems of native plants fix carbon better than shallow lawn grasses.

In cities, it's important to surround yourself with as much natural greenery as possible, says Mankiewicz, who is seeking funds to build rooftop greenhouses in New York City using lightweight soil. "The bigger the plants the better, because they invest a lot of energy in below-ground root systems so that more carbon can be fixed there."

If you live in the city, plant a rooftop garden. Even better, suggests Mankiewicz, build a greenhouse. There, you can shelter your plants from the elements, take advantage of a longer growing season, and get much larger plants to grow than you can in a rooftop garden. If you vent the greenhouse into your home and make sure it has "thermal mass" in, say, a concrete floor or a few strategically placed drums of water, you can also cut your fuel bills for home heating. (And you can tell your friends you're using a greenhouse to combat the greenhouse effect!)

Consumers and individuals can take any number of other actions in their daily lives to help solve the problem of warming. *Remember:* Americans lose as much energy through leaky windows and doors and other inefficiencies as flows through the Alaska pipeline.

Reducing fossil fuel use wherever possible is a good idea. Merely switching to hand tools won't do much to reduce carbon dioxide emissions, but buying a highly fuel-efficient car when your current car's days are over can make a difference, and so can heeding the other energy "shalt nots." Or, you can go further: become part of the solution. If you're building or remodeling a house, make it solar heated or at least energy efficient; if you live in an apartment, work with your landlord to retrofit your living space.

Become a citizen activist, for in this way you can make an even larger difference. Write to your legislators urging them to support the bills that advocate programs to prevent global warming; start local campaigns to save energy in your community; recycle.

If you want to set priorities, the most important step you can take as a consumer is to control fossil-fuel energy use in your car, home, and workplace. If you want an environmentally exemplary life style, be aware that there is even more to be done than recycling, tree planting, gardening, and taking some of the other small steps suggested here. You can, for instance, separate your toxic chemical trash (old paint, resins, weedkillers); avoid toxic chemical preparations in favor of naturally made ones; use solar products; and buy recycled paper products. If all of these ideas sound impossible, consider what Ronald Jager writes in his memoirs of World War II:

All the farms in our neighborhood had riches to contribute to the national harvest: First from the dump behind the house, then from the piles beyond the barn and beside the stone pile where lay many a broken hunk of cast iron and various abandoned implements of rusty steel. They had accumulated . . . through the years, and now they would be reborn as guns and tanks. . . .
After the heavy and rare metals came lighter stuff. School organizations put on scrap iron drives and paper drives and different school grades competed with each other. We brought packages of newspapers and old rusty pails full of bits of iron scrap to school on the bus and we threw the iron on the pile west of the school. Week by week,

the pile mounted higher, and as it rose, so did our pride in our country and in our school. We would keep those caissons rolling along—whatever caissons were. (I had no idea.)

When saving Earth from global warming becomes part of a national and international drive, then people will see how much they can accomplish as individuals. As Rep. Claudine Schneider told *The Environmental Law Reporter*, "Without a doubt, never before in history have individual actions come to play so great a role in the course of human events. Prevention pays, and it is incumbent upon each one of us to tap its benefits. Failure to do so will be a terrible squandering of our natural endowment."

6

Limits and Frontiers

On Spaceship Earth there are no passengers;
everybody is a member of the crew. We have
moved into an age in which everybody's
activities affect everybody else.

MARSHALL McLUHAN

For the first time in my life, I saw the horizon
as a curved line. It was accentuated by a thin
seam of dark blue light—our atmosphere.
Obviously, this was not the 'ocean' of air I
had been told it was so many times in my life.
I was terrified by its fragile appearance.

ULF MERBOLD
West German space shuttle astronaut

In a scene from a recent Star Trek movie, Captain Kirk awakes from
sleep while time traveling to find himself orbiting Earth. "Earth!" he
cries. "But when?" Faithful sidekick Spock checks the instrument
readings on the starship *Enterprise* to inform him, "Judging from the
pollution content of the atmosphere, I believe we have arrived at the
latter half of the twentieth century."

Science fiction? Real-life astronauts Captain Michael Coates and
Kathryn Sullivan recently returned from a series of NASA space shuttle
missions to report that the view of Earth from space is now clouded by
pollution. Shuttle photos showed industrial pollution blowing off the
coast of China, as well as over now industrializing Third World cities.
"When you fly into space . . . you get to see the earth as a finite entity,"
Coats says. "I think that everyone that's flown up there becomes much
more sensitive to environmental concerns than they were before they
flew."

It shouldn't take a spaceflight to convert the wary and the indifferent.
But Stanford University biologist Paul Ehrlich writes that we need to do
something similar from the ground, using state-of-the-art scientific

instruments to allow us to "fly blind"—that is, to accept the reality of problems that we cannot yet see or feel. Twenty years ago, in fact, some scientists, policy makers, and environmentalists began this "as if" journey while preparing for the landmark United Nations meeting in 1972 that created the U.N. Environment Programme. In this group's remarkably prescient report, *Only One Earth,* by René Dubos and Barbara Ward, they pointed out that "Our sudden, vast accelerations—in numbers, in the use of energy and new materials, in urbanization, in consumptive ideas, in consequent pollution—have set technological man on a course which could alter dangerously and perhaps irreversibly, the natural systems of his planet upon which his biological survival depends."

Twenty years since that prophetic statement, we technological beings have worsened our plight but have also made tremendous progress in understanding it. We've increased the load of carbon dioxide in the atmosphere by a little less than 10 percent, more than doubled the load of chlorofluorocarbons, and deforested the globe at an accelerated rate. In so doing, we have confirmed some of scientists' worst fears—that we would come up against the "limits to industrial growth." But, at the same time, international summit meetings now give as much attention to global ecology as to the global economy. NASA's proposed "Mission to Planet Earth" clearly acknowledges that understanding our own planet is as important as exploring new ones. And scientists are joining forces internationally for the first time to pool their knowledge about Earth's systems.

A LEADING ENVIRONMENTAL INDICATOR

Invisible to the naked eye, carbon dioxide is a kind of "leading environmental indicator" of planetary distress. If viewed century by century, the carbon dioxide content of the atmosphere can also be seen as a kind of index of the energy and speed of the civilizations that dwelled on the Earth.

The speed of technological change ushered in by the Industrial Revolution was a marked departure from the gradual evolution of human technology up until that point. For millennia, human beings relied on simple technology. As Dubos and Ward write, "Early civilizations were based, with few exceptions, on river valleys whose waters were managed so as to bring reliable water supplies to the farms. . . .

Management of waters required reliable measurements of land and flood and exact knowledge of times and seasons. Mathematics and astronomy were born among the Chaldeans and the Egyptians and later gave birth to the Greek vision of universal law which would embrace ultimate reality."

For another thousand years, they write, most of the technological instruments and tools of social organization then perfected stood humankind in good stead—alphabets, fire, metal smelting, boats that ran on wind power and knowledge of tides, and agricultural arts. "Napoleon's land armies went no faster than Hannibal's. Charcoal still smelted iron ore until the eighteenth century. Water wheels powered the first factories. The Arabs knew as much mathematics as Galileo."

The pace of social growth—population, energy sources, use of food and minerals—which began to grow in the seventeenth century, took off like a shot in the twentieth. Using estimates of population since the beginning of human history, the authors track world population as edging up to about 400 million by the fall of Rome. By 1600, the world had reached its first billion. Population growth accelerated in the nineteenth century as the Industrial Revolution picked up steam and infant mortality fell.

Today, world population is around 5 billion and is estimated to rise to a high of 12 to 14 billion by the middle of the next century. As earlier chapters of this book have detailed, fossil fuels production skyrocketed since the beginning of the twentieth century, accelerating most after World War II, until the first oil price shock of 1973. Today, fossil fuels are even more entrenched as the basis of our economy, and their use in the developing world is growing at an accelerating clip. The crisis that *Only One Earth* saw developing 20 years ago is indeed upon us.

During the last decade or so, the world's governments and its people have continued to deplete the planet's resources as though there were no tomorrow. Energy use, pollution, and world population have all grown, with the result that:

- Each year the planet's population grows by 90 million. That's the population equivalent of adding another Mexico or Japan *every year*.
- Human beings are coming more and more to dominate Earth's resources. According to Peter Vitousek, a Stanford University biologist

who heads the ecosystems advisory committee for the Organization for Tropical Studies, by 1986 human beings were using, diverting, or wasting 40 percent of all the nutrients and energy green plants accumulate. "For a single species to have that dominance is new, incredible, and probably unstable just in its own right," he concludes.

• Most of the pressure on natural resources is coming from the rich countries, where per capita energy use is highest. Half of two-thirds of the greenhouse gases injected into the atmosphere come from wealthy countries. The United States and the Soviet Union together account for almost a third of the total.

• U.S. energy consumption is the highest in the world. The latest data indicate that although U.S. citizens make up only about 5 percent of Earth's population, they use about one-fourth of the world's commercially traded fuel.

Some percentage of people have always been acquisitive and wasteful, but never before has their "conspicuous consumption" combined with the genuine needs of the far more numerous poor to threaten the planet's resources. And this increase isn't simply a numbers game. When E. F. Schumacher in 1973 voiced alarm at our overuse of fossil fuels and misuse of "the capital of living nature," he referred not only to the huge quantitative jump in our use of these fossil fuels at the time but also to "the unique qualitative jump" in our use of petrochemical-based compounds—the ability of science to fashion substances unknown to nature. "Against many of them, nature is virtually defenseless. There are no natural agents to attack them and break them down," he wrote. "It is as if aborigines were suddenly attacked with machine gun fire: Their bows and arrow are of no avail."

Schumacher was talking about chemicals in general and did not mention chlorofluorocarbons per se, but his point applies there as well. "It is only in the last twenty years or so that [new compounds] have made their appearance in bulk," he wrote. "Because they have no natural enemies, they tend to accumulate, and the long-term consequences of this accumulation are in many cases known to be extremely dangerous, and in other cases totally unpredictable." Such were his words 13 years before the ozone hole first appeared.

THE FORCES OF HABIT

The ozone hole can be understood as a metaphor for our own cognitive hole when it comes to facing our civilization's impact on our global home. A blind spot, writes Daniel Goleman in *Vital Signs, Simple Truths*, is "the gap in our field of vision that results from the architecture of the eye." What we perceive is a matter of how we select, what we filter out and allow in, he writes, "but the very capacity of the brain to do so makes it vulnerable to skewing what is admitted to awareness, what rejected." Breaking through people's denial systems is fundamental to solving the problem of the greenhouse effect—an already difficult task made more difficult by the unclear signals leaders send about whether rapid climate change poses a real threat.

Part of the problem is that few people want to confront a fact that implicitly calls for significant change. The famous line from Pogo— "We have met the enemy and it is us"—perfectly describes the roots of the greenhouse problem. As the editor of *New Perspectives Quarterly* puts it, "The evil is not extraordinary; it is banal. Our main enemy is not some bushy-browed Russian tank commander poised for invasion at the Fulda gap; it is the Los Angeles commuter driving to work."

Somewhat fatalistically, Paul Ehrlich and Robert Orenstein suggest that the human brain has not evolved enough to cope with gradual or long-term threats to the natural world. These authors write in *New World, New Mind* that human beings may not be "wired" to cope with invisible global stresses. Hundreds of thousands of years ago, they argue, our ancestors survived because they responded quickly to such threats as "a thrown spear" or "the darkening of the entrance to the cave as a giant bear enters." These are very different, say Ehrlich and Orenstein, from "threats like the slow atmospheric buildup of carbon dioxide." Our species, they say, is "mismatched" to the new perils of the world we've created, so "we are losing control of our future." The solution? "The human predicament requires a different kind of education and training to detect threats that materialize not in instants but in years or decades"—we need to develop "slow reflexes to supplement the quick ones" and self-consciously accept that "slow-motion crises" are indeed real and present dangers.

These scientists have a point. The time frame of the greenhouse problem is numbingly protracted. With a few possible exceptions, such as nuclear war, long-term plutonium storage, desertification of the North African lands that Rome salted, and species extinction, what other environmental problem in history has had the potential to span millennia without hope of self-correction? Geologists, paleontologists, and, recently, climatologists have been forced to visualize vast reaches of time, but for most of the rest of us, doing so is an act of pure imagination, not one of social responsibility.

Another theory is that we've *taught* ourselves not to respond to the new threats. San Francisco psychologist Katherine Burton believes that the "mobilization of a global will" will happen the same way transformations occur in cultures, families, and businesses. "We now live in global anarchy," Burton says. "The Alaska oil spill by Exxon was like the drunken dad getting drunk one night and coming home, saying, 'oh, I'm sorry.'" Society, she says, is much like the alcoholic family, angry at dad, but basically colluding with his behavior rather than upsetting the applecart. "The only way that the cycle is broken is to have a family conference—we're not going to allow this anymore—the game is up."

However influential such inborn and psychological barriers may be, they are by no means the only obstacles to action. Despite the scientific consensus that a continuing buildup of carbon dioxide and other gases will ultimately lead to a global warming, the public and its leaders have been treated to some potentially confusing information and theories while that consensus was forming. As *The Economist* magazine of London notes in a recent article on rapid climate change, "it is in the nature of public debates about scientific controversies that the alarms are followed by a bout of debunking."

Still another problem is that we're doing too little, too late. Twenty years ago, the United States passed a series of landmark environmental laws to clean the air and the water and control hazardous waste, solid waste, and toxic chemicals. But these laws went only so far. They've reduced the damage and diluted the strength of pollutants but haven't really stopped their flow. As a result, many of America's groundwater supplies are contaminated or threatened by pollution, and many of its cities don't meet safe air standards. Only now are experts realizing that we've created a kind of shell game in which toxic chemicals removed from the air, water, or soil inevitably end up appearing in one of the other two. Chemicals prevented from entering lakes and rivers, for example, end up in landfills that then leak into groundwater later.

*"This past summer, I got deeply depressed about our planet—as
if I didn't have enough problems of my own."*

This realization has spurred some recent efforts to reduce pollutants
or even eliminate them at the source. But the piecemeal approach is still
prevalent. "With some justification, environmentalists have gone for
obvious targets such as vehicle emissions, power plant pollution, the
aerosol link and the ozone layer, offshore incineration of toxic wastes,
dumping at sea, nuclear power and such bestialities as modern whal-
ing," writes the coeditor of the British magazine *The Ecologist*. "But
with rare exceptions, environmental groups have avoided looking at the
whole industrial enterprise."

Just as most people anaesthetize themselves against the nuclear
threat, so too, they avoid facing global environmental issues like ozone
depletion or climate change until a crisis or a disaster hits. "When
there's a heat wave," climatologist Stephen Schneider told a panel at
Stanford University, "the phone rings. Increasingly, there have been
larger numbers of phone ringings." Another scientist, Peter Raven,
director of the Missouri Botanical Garden, thinks the problem stems
from people's failure to think globally. "As long as most of us are living
in comfortable surroundings in the United States, it's difficult to realize

[how local action and the global environment are] interwoven," he says.

So the average person gets the ozone hole mixed up with ground-level ozone, and some get global warming confused with increased ultraviolet radiation. This gives rise to a kind of gallows humor in which problems are so many that one is suggested as a solution for another—"Let's take the ground-level ozone and patch up the sky," or "Let's take all the greenhouse gases and shoot them through the ozone hole." It's tempting for even socially concerned people to throw up their hands and leave such complicated issues to the experts.

A final impediment to action on the greenhouse problem is that some of the work needed to solve it is too routine. Robert Stobaugh and Daniel Yergin have written that one of the main obstacles to energy efficiency—a key defense against global warming—is its lack of glamour. "There is in this country a natural-enough desire for a technological fix, preferably one big one that will solve all energy problems—another Manhattan Project, another man-in-space program." Indeed, some survey data indicate that many Americans simply assume that high technology will step in to "save" them just in time. "Conservation," write Stobaugh and Yergin, "is prosaic, even boring." By the same token, "source reduction" or waste minimization isn't an issue that can fire peoples' hearts.

In the face of such obstacles, what will it take to spur action on the related threats of climate change, ozone depletion, and global resource depletion? And given inertia, can government fast-track programs to keep up with the speed of ecological change? To answer these questions for yourself, consider what has happened to the environmental movement that began two decades ago.

TOWARD AN ECOLOGICAL ETHIC

Many of the issues that global warming now raises about energy overconsumption and the "throwaway society" more generally were first raised in the 1960s and 1970s. The Club of Rome announced that civilization had embarked on an era of limits, and ecological perspectives touched many for the first time. People began to question what industrial society has brought, and some—mostly the young—revolted against consumerist ideals.

During these same decades, E.F. Schumacher, Robin Clarke, Victor

Papanek, Amory Lovins, and other visionary critics of latter-day industrial society charted a new path for technology that was largely small-scale, inexpensive, accessible, compatible with human needs, and ecologically responsible. Between the late 1970s and early 1980s solar, wind, and other alternative energy sources began to make some strides. But this countercurrent never swelled large enough to turn the mainstream, and few in the business world ever took "small is beautiful" seriously. The higher energy prices that plagued consumers during the late 1970s, which tended to make people want to conserve, were brought under control first by worldwide recession in the early 1980s and then by the return in 1987 of cheap oil. By then government support for the New Age tinkerers had also eroded.

Membership in environmental groups swelled throughout the 1970s—to some 10 million today, according to *American Demographics* magazine—and the late 1970s saw new environmental groups grow and subdivide like hot stocks, especially in response to unpopular political appointments and environmentally damaging accidents. But many adults who came of age when the environmental movement of the 1970s crested now have children, cars, houses, boats, and closets and garages full of high-tech gadgetry. And this group has proved just as susceptible to advertising as any other "target group," segmented by market researchers into such categories as "Belongers," "Achievers," or "Inner-Directeds." As Langdon Winner writes in *The Whale and the Reactor: A Search for Limits in an Age of High Technology*, "Rather than attempt to change the structures that vexed them, young Americans growing older were settling for exquisite palliatives." Indeed, did life styles really change, or simply a generation's fashions?

For reasons still debated, Americans' interest in energy conservation and the environment declined in the 1980s. Memories of the oil crisis were fading, oil prices fell and remained stable, government took a misguided (and ultimately unpopular) approach to many energy and environmental issues, the baby boomers were aging, and their children were exercising each generation's right to rebel against their parents' values. Even larger historical forces may have been in play, too. "In essence, the 1960s hastened the erosion of the false and devastating universal ideal of 'progress' embodied in the pyramidal Trinity of Father, Science, and State—a faith that had underlain our assumptions that the world displays linear order, historical sequence, and moral clarities," sociologist Todd Gitlin reflects in *Postmodernism: Roots and Politics*. "A new moral structure has not yet been built and our culture has not yet found a language for articulating the new understandings we are trying, haltingly, to live with. It objects to all principles, all commitments, all crusades—in the name of unconscientious evasion."

As we enter the 1990s, questioning values is becoming a popular pastime, says congressional aide Mike Totten, who works for Rhode Island Republican Claudine Schneider—a strong proponent of renewable energy and conservation programs. "As the decade draws to a close, I hear more talk, countless conversations on [Capitol] Hill, about values and how they've been shaken, and even cardinal sins. People talk about hubris in government and education; greed in the corporate world as well as in Congress with its PAC money . . . sloth in the public—the feeling of no matter how bad it is, I don't want to hear about it. All in the face of problems getting so much more insurmountable." But, says Totten, "more and more I get a sense that there's another shift among people occurring, maybe not so widescale as the movements of the 1960s and 1970s, and people are talking about replacing hubris with humility, sloth with stewardship, in the practical real-world knowledge sense of stewardship and charity instead of greed."

Totten may be right. The mid to late 1980s saw a rash of books on the subject of values—material and otherwise. The titles themselves are a commentary on the end of the "me decade": Paul Wachtel's *The Poverty of Affluence: Individualism and Commitment in American Life*; Robert N. Bellah and colleagues' *Habits of the Heart*; Laurence Shames' *The Hunger for More: The Search for Values in an Age of Greed*; Lewis Lapham's *Money and Class in America: Notes and Observations on America's Civil Religion*; and more. By the late 1980s, the market for environmental books and "new age" books on such subjects as the Gaia hypothesis also revived after a decade-long slump.

Book sales do not a revolution make, but they do indicate that some people are beginning to question the "shop till you drop" ethic and the connection between standard of living and quality of life. "Like everything in TV-oriented culture, it's a matter of the pendulum swinging one way or another," says Denis Hayes, founder of Earth Day and organizer of Earth Day 1990. "And the pendulum seems to be moving away from the lifestyle of the Rich and Famous. If the pendulum theory is correct, we are poised to achieve the best possible time for a movement toward more public-spiritedness and less self-interest."

Now, suddenly, nature is fashionable, rain forests are in. Record companies are inserting information packets on biodiversity and tropical hardwoods. Rock musicians such as Sting and R.E.M. and Bono are writing songs about saving the planet.

Some environmentalists are counting on average people to emulate celebrities' renewed commitment to the environmental cause. But regardless of what famous people do, the rest of us have to be concerned about more than ourselves if we want to halt rapid climate change. And we have to recognize our role in the problem. Denis Hayes believes that greater concern about the fate of the planet is already beginning to strengthen spiritual and religious values. According to Hayes, environmental events—tree-planting efforts and the like—will more and more trigger people's "concern for husbanding the earth in a spiritual way." "If you get people out in nature long enough, people want to care for it," he claims.

Indeed, one locus of change seems to be the churches, where the "ecological stewardship" or Christian ecology movement is gaining force. The churches see the greenhouse effect and ozone depletion as a case of mounting abuse of God's creation. In a preface to a Presbyterian resource paper, "Keeping and Healing the Creation," theologian Robert Stivers of Pacific Lutheran University laments that the church's entry into this debate may already be too late. "Perceptions of something radically wrong in the natural environment first came to the attention of the church in the 1960s. The 1970s saw a crescendo of activity . . . [but now] a new birth of enthusiasm and reflection is badly needed." Says Stivers, the biblical injunction to "till" has taken precedence over the command to "keep." In a similar vein, Presbyterian minister William Gibson has written that "In struggling to survive, in seeking to live ever more comfortably, and in failing to distribute production equitably, we humans have made demands that now increasingly exceed what nature can sustain."

Described by writer Pat Stone in *Mother Earth News*, the eco-justice or "Christian ecology" movement is infiltrating many Christian denominations and taking in Jews and people of other faiths as well. "Quakers now publish environmental manifestoes. Evangelist Billy Graham pronounces that 'we Christians have a responsibility to take a lead in trying to take care of the Earth.' Pastors build solar and superinsulated churches to be better 'stewards of God's resources.' " And theologians of various faiths, Stone writes, are all trying to prove their own religion's environmental heritage in scripture and dogma.

Some of the churches are reaching outside the fold to make converts to ecology. Stone points to signs that Christian ecology is "on the verge of becoming a national force." She cites Wall Street megatrend analyst Dan Blum's forecast that "church-led environmentalism will become a major trend in the 1990s." And she argues that just as the American Revolution was preached from pulpits and civil rights issues pushed by priests and rabbis, organized religions "though slow to stir, *can* make a mighty impact when wakened into action."

In a foreword to a new edition of *Only One Earth*, Norwegian prime minister Gro Harlem Brundtland writes, "The problems we face as a world community are planetary, but not insoluble. Our two greatest resources, land and people, can still redeem the promise of development. If we take care of nature, nature will take care of us. But the huge changes sweeping over us and our biosphere demand fundamental changes in our attitudes, our policies and in the way we run our societies." Brundtland's message is on target, and it also seems to indicate that some political and church leaders are thinking along the same lines—surely another sign of hope that the environmental movement isn't dead and that we can do something about global warming before it is too late.

TECHNOLOGY THAT MIMICS NATURE

As Barry Commoner pointed out two decades ago in *The Closing Circle*, environmental problems weren't so bad in most of the world until after World War II even though we enjoyed rapid technological progress before then. But something changed. The scale and nature of human activity shifted, and now pollution taints the atmosphere and threatens the systems that control climate and sustain life on the planet.

This hard fact suggests that while economic progress and environmental protection aren't necessarily incompatible, a technological transformation is needed if we are to pursue both. World Resources Institute president Gus Speth argues that we must rethink our use of twentieth-century technologies and "rapidly adopt different, environmentally sustainable technologies for the twenty-first century." Pollution *control* may have bought us time but now we have run out. As a result, says Speth, "We have done all we can with end-of-the-pipe solutions; now environmental goals must be built into industrial and agricultural design." If we want to keep the expense and dislocation of a technological shift manageable, we should take the first steps now.

The threat of global warming makes the need for renewable energy far more urgent than it was back in the 1970s. And it requires replacing "dirty tech" with "clean tech"—abandoning old manufacturing processes for closed-loop systems that minimize or eliminate waste instead of dumping it into the environment.

Many renewable-energy technologies, including wind and water power, have been harnessed since the times of the ancients. But capturing renewable energy and imitating nature in other ways requires mastering complex physical principles. Ingeniously designed rotor airfoils, tail vanes, and thin-film solar cells are veritable engineering marvels. To be sure, dreams of technological simplicity may have to be abandoned if we truly accept the call to "design with nature." As one 1972 ecology text, *Ecology, Pollution, Environment* puts it: "There is no machine that can match living organisms in complexity and diversity. Animals and plants, unlike machines, can feed and repair themselves, adjust to new external influences, and reproduce themselves. These abilities depend on very complex interrelationships among the separate parts of the body. Thus, each of us human beings is far more than a sum of brain, heart, liver, stomach, and other organs. What affects one part of the body affects all. Each of us is thus a living *system* of interdependent parts. The system functions in an extremely complex manner, so complex that it is far from fully understood."

To make headway against rapid climate change, the United States and other industrialized countries need to design ecological principles into industrial processes rather than treat pollution as an afterthought. "In the future, environmental protection must increasingly become a *pro-active*, affirmative process of designing the major sectors of economic activity—manufacturing, housing, agriculture, transportation, energy—so that they fulfill economic needs without destroying regional and global environments," argues Speth. Environmental protection, in other words, ought to move from regulation to redesign.

The last serious, broad-based attempt to redesign technology was in the 1970s. Experiments then with "appropriate technology," "soft tech," and "alternative technology"—technology designed to replace overbuilt and pollution-prone machinery and systems—ultimately missed their mark. As Langdon Winner writes, most proponents of "A.T." ignored "the facts of organized social and political power" and the history of modern technology. Yet, says Winner, this movement has had an important and lasting influence on how people think about technology. It challenged modern technical orthodoxy, broadened the meaning of such concepts as efficiency and sustainability, and got engineers, manufacturers, and consumers to think twice about these concepts.

However exaggerated other claims for it might be, alternative technology has also made a practical difference—a quantifiable dent in our dependence on fossil fuels. Already, billions of kilowatt-hours of power are being supplied by direct solar energy, geothermal, wind, photovoltaic cells, biomass, and hydropower. Although the lion's share of this sizable contribution (12 percent of domestic energy supply) comes from hydropower produced in large dams, a significant share is also being harnessed by alternative technology.

As for what technology is appropriate in the energy transition we now face, it seems clear that it will have to build on and improve the best engineering concepts of recent decades, remain in sync with ecological principles, and reflect the economic and political facts of life ignored the last time we took a serious look at our society's technological base.

THE BEAUTY OF THE FIX

There's no aspirin for our planet's fever. The only way to control the greenhouse problem is to cut fossil fuel emissions. But there's beauty in that fix. Not only will it slow down the greenhouse effect; it could also help conquer many other major environmental ills—from acid rain and other forms of air and water pollution to waste and pesticide contamination. At the same time, these measures can contribute to such national objectives as energy independence, sustainable agriculture, international competitiveness, national security and strategic-materials preservation, and economic stability.

Fossil fuel use is a common element in all of these complex, interrelated issues. Consider just a few of the links:

- *Air pollution.* The combustion of coal, oil, and gas releases sulfur dioxide and other pollutants that eventually form acid rain and ground-level pollution (smog).
- *Toxic Pollution.* Oil is the basis for petrochemical-based pesticides, which often remain as toxic residues on our food, as drift in the air, and as contaminants in rivers and streams.
- *Energy Insecurity.* By depending on more and more imported oil, the United States puts its national security and economic stability at greater risk. Economically, imported oil is a major contributor to the United States' swollen trade deficit. And, as a World Resources Institute report notes, "we are more dependent on imports now than we were in 1973 when history showed us how much a disruption could hurt." Although we could tap into the Strategic Petroleum Reserve in an emergency, that reserve wouldn't keep OPEC from raising oil prices once it has reestablished control over world markets.

Fortunately, the multiple benefits of switching away from fossil fuels have been amply assessed. And a wealth of interesting new information is providing insights into how new practices and technologies can be forged. Traditional technologies and practices abandoned as fossil fuels slowly saturated our economy are also being shown to deserve a second try.

One such rediscovery is sustainable or "regenerative" agriculture. In some of its forms, farmers switch from petrochemical-based weed-killers and fertilizers to more "natural" methods, such as rotating crops and using organic mulches and other materials to fertilize crops and control pests. This scattered movement—defined by Robert Rodale, one of its leading proponents, as "farming and gardening systems that regenerate rather than erode the productivity of natural resources"—is now gaining ground across the farm belt. "As the sustainable agriculture movement gains momentum, some experts are predicting as much as a 30-percent reduction in farm-chemical usage," the *Wall Street Journal* reported in May 1989. Moving away from energy-intensive agriculture produces many benefits. "If you use less fertilizer you save money and reduce groundwater contamination," says Nick Sundt of the U.S. Office of Technology Assessment. "If you use less pesticides, you reduce pollution and worker diseases; if you use drought-resistant crops, you save water, and so on." Rodale and others also claim that this approach protects and restores farmland.

Such discoveries and rediscoveries bring hope, but none too soon. If we continue to emit carbon dioxide at the current rate, we are headed for enormous economic setbacks and unknown ecological shifts. "The Los Angelesization of the planet," writes cultural historian William Irwin Thompson, "cannot take place, for in the greenhouse effect nature has her own negative feedback mechanisms for shutting down the furnace of industrial civilization."

Facing this prospect, the true cynic might say that it would be better for the planet if we were not here: "Maybe Earth needs a rest," Spalding Gray concludes in his eccentric movie, *Swimming to Cambodia*. Indeed, if scientist James Lovelock's hypothesis—that Earth is a self-regulating organism that will respond to disturbances by shifting into a new stable state that human beings might not find habitable—proves correct, and if we don't take the threat of a rapid global warming seriously, it may be millions of chortling bacteria that have the last laugh on us.

But if we want our planet's climate to stay pretty much as it has been throughout human history, our perspectives have to change and so do our leaders' perspectives. As one leader, Gro Harlem Brundtland puts it, in the face of rapid environmental deterioration, sustained cultural change "no longer seems utopian."

Perhaps more than any other environmental issue, the prospect of climate change reminds us that Earth is both one of a kind and finite. Dorothy preferred it to the magical Oz, and the astronaut Michael Coates said after his last mission that "You get this feeling like here's this little planet earth where all of us live going through this huge black void. If we mess up this planet, there's no place else to go. This is it—we've been blessed with a very beautiful place to live and we kind of take it for granted."

APPENDIX 1

Two Futures

To test the effects of various policy strategies, the World Resources Institute developed a computer model to link information about population growth, resource prices, technology developments, and consumer choices to future rates of growth in energy use and greenhouse gas emissions. These projections are used to estimate the ultimate warming effects caused by various atmospheric changes. No model can predict future greenhouse gas emissions with any certainty because these emissions depend on so many economic decisions not yet made and, possibly, on quirks of nature not yet revealed. But this model does show that, while it's too late to prevent all future global warming, there is still time to avoid "worst case" catastrophes.

BUSINESS AS USUAL

If past is prologue, most policy makers will want to do as little to bring climate change under control as conscience and public pressure permit. As Larry Tye of the *Boston Globe* puts it, too often "world leaders are taking their cues from the Wizard of Oz in tackling global warming and other pollution perils: They loudly proclaim their resolve to act, reassure a nervous public with dramatic displays in huge halls and wait for the problems to go away." The high costs of this approach to agriculture, industry, coastal regions, and the human body ought to rouse our leaders, but what if they don't? What might happen?

Consider what could happen if no significant action is taken to slow global warming. Let's assume that energy use continues to grow at the same rate it did during the 1980s (about 2 percent per year) for the next 35 years. Assume also that tropical deforestation continues, but the rate declines over time as nations begin to see the economic folly of selling a source of long-term wealth to turn quick profits. All countries sign the Montreal Protocol to protect the ozone layer—not a radical move since the political momentum is already gathering and the control measures are relatively simple. But governments don't go any further toward reducing CFC production than this international agreement demands: a 50 percent cut for industrialized countries by 1999 and a two-thirds pound–per capita limit for countries that are not now major users.

Maintaining the status quo would mean that life styles wouldn't change much until climate did (except, of course, for unrelated reasons). New energy technologies might or might not be introduced, depending on the marketplace. Per capita energy consumption would grow by almost 50 percent in the next 35 years, and economic growth would be only slightly slower than it has been over the past few decades.

This do-nothing approach, our computer model informs us, commits Earth by 2020 to a total warming of 1.8 to 5.4 degrees Fahrenheit above what temperatures were, on average, a century ago. By 2075, the heat will really be on: the planet is committed to a warming of between 5.8 and 17.5 degrees Fahrenheit, relative to preindustrial temperatures. The concentration of carbon dioxide in the atmosphere will reach over 700 parts per million, compared to 350 parts per million today.

RISING TO THE OCCASION

What if governments all over the world decide to do everything they could in the next few years to stabilize global climate as soon as possible? The decision itself may be half the battle, but only half. These governments would have to back their commitments with sweeping policy changes. As the *Washington Post* notes, "Any greenhouse agreement worth its salt is going to require major changes in the U.S. economy and lifestyle."

The first such major change must be to improve energy efficiency dramatically—squeezing more work out of every dollar spent on energy. The industrialized countries have a head start, since more efficient technology is available and in some cases already in use in these nations. In this scenario, they would have to reduce the amount of energy used per constant dollar of gross national product by about 1.4 to 2.4 percent a year, on average, for 40 years. To support continued economic growth, developing countries would presumably introduce new technologies that rapidly increase the economic efficiency of purchased energy—by about 1.7 to 2.4 percent per year—during those four decades.

Step two, if we rise to the occasion, is to make average fuel prices reflect the environmental costs of using energy. In this scenario, the fuels that release the most carbon dioxide per unit of energy (coal and oil shale, for instance) would be taxed at a higher rate than cleaner-burning fuels. At the same time, the fuel-cycle security, plant-decommissioning, insurance, and waste-disposal costs of nuclear fission would be incorporated into the price of electricity generated by nuclear power plants. That way, individual consumers and corporations would know the full costs of energy supply and use when they make investment choices.

Rising to the occasion also means rapidly phasing out chlorofluorocarbons (CFCs) and not bringing in substitutes that cause the same problems. Besides signing the Montreal Protocol, all nations would soon after the turn of the century also take other steps to control ozone-depleting chemicals. They would rapidly develop CFC substitutes that don't contribute at all to global warming or ozone depletion. (Aerosols might be propelled by nitrogen gas. Foams could be blown with carbon dioxide and refrigerators run on ammonia or helium.) By 2010, the industrialized countries would have just about kicked the CFC habit, phasing out all but the 5 percent needed for medical and other essential purposes. The replacements would be limited to compounds that have no ozone-depletion potential or global warming effect. By 2020, CFCs and like compounds would be a thing of the past in industrialized countries. By 2050, the developing countries would also have left them behind.

The human side of this scenario is population growth. Through public education, better health care, and other measures, global population would be held to 8 billion people in 2075—3 billion more than now inhabit the planet but 6 billion fewer than today's highest expert projection for that time. Rising incomes and increasing services would presumably prompt people to feel less compelled to have many children to provide for them in their old age. If all other factors were equal, keeping population growth within these bounds would give us about one extra decade before Earth's temperature increased by twice as much as it has since the greenhouse problem began during the Industrial Revolution. (This grace period would be longer if population growth rates were the same in all countries. But they aren't: the nations where population is growing the fastest release the least CFCs per capita.) On the other hand, if "worst case" population projections come true, keeping greenhouse emissions down will require draconian measures that ill-suit any free-spirited people.

If governments pull out all stops to stabilize the climate, as described here, the payoff would be tremendous. By 2075, temperatures would still range from 1.4 to 3.1 Fahrenheit above today's temperatures; the total increase since preindustrial times would be about 3 to 7 degrees Fahrenheit. But these increases would come mostly from greenhouse gases released *before 1988*. After 2050, there is no additional commitment to future global warming. By about 2075, we are home free: temperatures could stop rising.

APPENDIX 2

GREENHOUSE POLITICS: A TIMELINE

1979

June
Report to President's Council on Environmental Quality by George Woodwell of the Woods Hole Biological Laboratory, Gordon MacDonald of Mitre Corporation, and David Keeling and Roger Revelle of the Scripps Institute of Oceanography: "Man is setting in motion a series of events that seem certain to cause a significant warming of world climate unless mitigating steps are taken immediately."

October
National Academy of Sciences study on the greenhouse effect initiated by President Carter's science advisor warns that a doubling of carbon dioxide in the air would raise global temperatures by 2.7 to 8.1 degrees Fahrenheit.

1980

January
Council on Environmental Quality chairman Gus Speth releases a report urging inclusion of the carbon dioxide problem in U.S. and global energy policies.
Scientists at the Goddard Institute of Space Studies highlight in a report the greenhouse gas contribution of methane, ozone, nitrous oxide, and chlorofluorocarbons (CFCs).

1983

An Environmental Protection Agency (EPA) report, "Can We Delay a Greenhouse Warming?" concludes that, despite major policy initiatives, current carbon dioxide emissions are enough to *raise* temperatures 3.6 degrees Fahrenheit.
A National Academy of Sciences greenhouse study, "Changing Climate," states that current evidence is not enough to warrant changes in energy policy.

1985

Atmospheric scientists V. Ramanathan of the University of Chicago and Ralph Cicerone from the National Center for Atmospheric Research find that all the other greenhouse gases combined trap as much heat per decade as carbon dioxide does.

October
Scientists from 29 nations convene in Villach, Austria, under the auspices of the United Nations Environment Programme and the World Meteorological Organization: "Some warming . . . now appears inevitable."

December
Sen. Albert Gore calls for an "International Year of the Greenhouse."

1986

January
The World Meteorological Organization and the National Aeronautics and Space Administration (NASA) issue a three-volume report on atmospheric ozone.

June
At Senate hearings, James Hansen from NASA's Goddard Institute of Space Studies predicts a U.S. warming of 3 degrees Fahrenheit over 30 years if current levels of carbon dioxide emissions continue. NASA presents evidence of an ozone hole over Antarctica at the same hearing. Sen. John Chafee asks EPA and the Office of Technology Assessment to develop policy options for stabilizing greenhouse gas concentrations in the atmosphere. Congress earmarks extra money in EPA's budget to enhance climate change research.

November
EPA Administrator Lee Thomas says that the United States will propose an international phaseout of CFCs.

1987

January

V. Ramanathan testifies before the Senate Environment and Public Works Committee that Earth is already committed to a 1.3- to 3.6-degree Fahrenheit warming. Wallace Broecker of the Lamont-Dougherty Geological Observatory asserts that greenhouse warming could lead to large and unpredicted changes in global climate.

September

Twenty-four nations sign the "Montreal Protocol on Substances that Deplete the Ozone Layer," which calls for CFC consumption to be cut in half by 1999.

November

At a Senate Energy Committee hearing, Gordon MacDonald proposes a "carbon tax" to reduce fossil fuel combustion. Gus Speth of the World Resources Institute urges Congress to pass comprehensive legislation to combat global warming.

December

President Reagan and General Secretary Gorbachev issue a communique from the U.S./USSR summit to continue joint studies on global climate change.

1988

January

President Reagan signs the Global Climate Protection Act, requiring the President to propose policy responses to Congress.

A U.S./USSR working group is set up to develop "response strategies" to the greenhouse effect.

April

A United Nations Environment Programme/World Meteorological Association report, "Development Policies for Responding to Climate Change," says climate change will outpace natural systems' ability to respond and adapt.

The United States ratifies the Montreal Protocol, which limits production and use of the most common CFCs to 50 percent of the 1986 level in industrial countries. NASA and National Oceanic and Atmospheric Administration scientists on the U.S. Ozone Trends Panel confirm reports of Arctic ozone loss and declare CFCs the main culprit behind the Antarctic ozone hole.

June
James Hansen testifies before a Senate committee: "Global warming is now large enough that we can ascribe with a high degree of confidence a cause and effect relationship to the greenhouse effect."
At an international conference in Toronto, "The Changing Atmosphere: Implications for Global Security," the Prime Minister of Norway, Gro Harlem Brundtland, calls for a global convention on the greenhouse effect. The conference statement requests a 20 percent cut of carbon dioxide emissions by 2005.

August
Candidate George Bush states that "Those who are worried about the greenhouse effect are ignoring the White House Effect."

November
An Intergovernmental Panel on Climate Change is established in Geneva. The United States agrees to chair the Response Strategies Working Group.

1989

January
The British Meteorological Office announces that 1988 is the warmest year in the century-long temperature record. The British also state that the six warmest years in the temperature record have been in the 1980s. British scientists explain that the temperature rise is "consistent" with the greenhouse effect but "no unambiguous connection can yet be made."

March
British Prime Minister Margaret Thatcher leads a conference on the ozone layer for delegates from more than one hundred countries. In the

days before the conference, the European Community and the United States call for strengthening the Montreal Protocol and phasing out CFCs.

April
British Prime Minister Margaret Thatcher holds an all-day session with her cabinet and with scientists on the greenhouse effect.

May
President Bush announces his support for the negotiation of a climate convention, as United States negotiators in Geneva invite delegates of the Intergovernmental Panel on Climate Change to a major workshop in Washington to discuss the elements of such a convention. The British government announces its support for the negotiation of a climate convention.

The Helsinki Declaration: at the United Nations Environment Programme meeting on the ozone layer, delegates from more than 81 nations agree that the Montreal Protocol should be revised and CFCs phased out.

The United Nations Environment Programme Governing Council meets in Nairobi. Climate change is a major issue. The Governing Council resolves that negotiations for a climate change convention should begin immediately after a report by the Intergovernmental Panel on Climate Change is submitted to international bodies in late 1989. The Governing Council urges governments to consider options for reducing greenhouse gases while negotiations proceed.

July
Paris, France: At an economic summit meeting, 15 heads of state declare that the "conclusion" of a framework is urgently required and "strongly advocate common efforts to limit emissions of carbon dioxide and other greenhouse gases."

September
The Japanese government hosts a World Conference for Global Environment and Human Response toward Sustainable Development, focusing on climate change and deforestation.

October

IPCC delegates discuss elements of a framework convention on climate at Geneva meeting.

November

The Dutch government holds the first ministerial meeting on global warming. Sixty countries attend.

Industrial nations agree that carbon dioxide emissions should be stabilized as soon as possible.

Dutch agree to lower their own carbon-dioxide emissions.

December

At the Malta Summit President Bush proposes that the U.S. host the first negotiating session on a Climate Convention in the fall of 1990. The president also proposes a White House conference on global warming in the spring.

Source: Rafe Pomerance, World Resources Institute

APPENDIX **3**

SOURCES FOR THE RESOURCEFUL
Energy Conservation

The Most Energy Efficient Appliances. Published by American Council for an Energy-Efficient Economy (ACEEE). Annual ratings on the energy efficiency of residential equipment on the market. ACEEE, 1001 Connecticut Ave., NW, Suite 535, Washington, DC 20036. Cost: $2.00.

Saving Energy and Money with Home Appliances. Produced by the Environmental Science Department of the Massachusetts Audubon Society and the American Council for an Energy-Efficient Economy (ACEEE). A 34-page guide to buying and using energy-efficient appliances: water heaters, refrigerators and freezers, air conditioners and heat pumps, stoves, washers and dryers, dishwashers, portable space heaters, and lighting. ACEEE, 1001 Connecticut Ave., NW, Suite 535, Washington, DC 20036. Cost: $2.00.

Gardening

Native Shrubs for Landscaping. By Sally Taylor et al. Guide to basic landscape design principles, including how to obtain native shrubs and the planting and care of shrubs native to the East Coast. The Connecticut College Arboretum, Connecticut College, New London, CT 06320. Cost: $3.50, plus .70 postage.

Energy Conservation on the Home Grounds: The Role of Naturalistic Landscaping. 1975. By William Niering. How to convert your home lawn or garden into a natural greenspace—for example, turning your lawn into wild grassland or landscaping using rocks and shade. The Connecticut College Arboretum, Connecticut College, New London, CT 06320. Cost: $1.00, plus $0.70 postage.

Tree Planting

The Global Releaf Report and *Citizen Action Guide.* Published by the American Forestry Association. The report is an environmental magazine published six times a year, and the guide is a how-to for preserving

your local environment. American Forestry Association, P.O. Box
2000, Washington, DC 20013. Cost: $25.00 donation.

Step-by-Step Guide to Organizing a Releaf Effort. Published by Trust
for Public Land. A fact sheet on launching a tree-planting campaign,
including how to identify available planting spaces. Trust for Public
Land, 116 New Montgomery, 4th Floor, San Francisco, CA 94105.
Cost: FREE.

Products

The Green Consumer. By John Elkington, Julia Hailes, and Joel
Makower. A comprehensive guide to environmentally safe consumer
products ranging from kitchen appliances to automobiles. Penguin
Books, 40 W. 23rd Street, New York, NY 10010. Price: $8.95.

Seventh Generation Catalog. Published by Seventh Generation. A
selection of environmentally sensitive products, from recycled paper
products to solar cookbooks. Seventh Generation, Dept. No. P089103,
10 Farrell St., South Burlington, VT 05403. Cost: $2.00.

*Shopping for a Better World: A Quick and Easy Guide to Socially
Responsible Supermarket Shopping.* Council on Economic Priorities,
30 Irving Pl., New York, NY 10003. Cost: $5.95.

Atmosphere. Published by Friends of the Earth. Quarterly newsletter
reporting on ozone protection activities around the world. Friends of the
Earth, 218 D St., SE, Washington, DC 20003. Cost: $15.00 per year
for nonmembers, $25.00 per year for corporate subscriptions.

Reducing the Rate of Global Warming: The States' Role. Published
by Renew America. State-by-state breakdown of carbon dioxide emissions from the burning of coal, oil, and natural gas; outlines efforts in
various states to help reduce buildup of these gases. Renew America,
Suite 710, 1400 16th St., NW, Washington, DC 20036. Cost: $10.00.

Facts Sheet on Vehicle Efficiency. Energy Conservation Coalition,
1525 New Hampshire Ave., NW, Washington, DC 20036. Cost:
FREE.

Community Recycling

*Waste to Wealth: A Business Guide for Community Recycling Enter-
prises.* By Neil Seldman and Jon Huls. A guide to investing in recycling

businesses. Institute for Local Self-Reliance, 2425 18th St., NW, Washington, DC 20009. Cost: $50.00, plus $2.00 postage.

Kilowatt Counter. By Gil Friend and David Morris. Describes energy and how it is measured; basic methods for conserving energy; energy use required per year for most household appliances; alternative energy sources. Institute for Local Self-Reliance, 2425 18th St., NW, Washington, DC 20009. Cost: $8.95, plus $2.00 postage.

A Guide to Recycling in Your Community. Published by the Institute for Local Self-Reliance. How to set up a residential waste recovery program: deciding what materials to collect; promoting the recycling program; implementing drop-off and pickup services. Institute for Local Self-Reliance, 2425 18th St., NW, Washington, DC 20009. Cost: $5.00, plus $2.00 postage.

Salvaging the Future: Waste-Based Production. Published by the Institute for Local Self-Reliance. Outlines paper, glass, plastics, and aluminum recycling for an area of one million inhabitants: the supply and demand for each material and the technology associated with its recycling; the cost of establishing scrap-based manufacturing plants; ways to devise "closed-loop" manufacturing in your community. Institute for Local Self-Reliance, 2425 18th St., NW, Washington, DC 20009. Cost: $100.00 (discount price for community groups: $35.00), plus $2.00 postage.

APPENDIX 4

ORGANIZATIONS THAT CAN HELP

In the United States

American Forestry Association
(and Global Releaf Project)
P.O. Box 2000
Washington, DC 20013

Center for Environmental
Information
33 South Washington St.
Rochester, NY 14608

Climate Institute
316 Pennsylvania Ave., SE
Suite 403
Washington, DC 20003

Environmental Defense Fund
1616 P St., NW, Suite 150
Washington, DC 20036

Environmental Policy Institute
218 D St., SE
Washington, DC 20003

Environmental Protection Agency
Public Information Center
401 M St., NW
Washington, D.C. 20460

Friends of the Earth
218 D St., SE
Washington, DC 20003

National Aeronautics and Space
Administration
Public Services
400 Maryland Ave., SW
Washington, DC 20546
(202) 453-8400

National Audubon Society
801 Pennsylvania Ave., SE
No. 301
Washington, DC 20003

National Oceanographic and
Atmospheric Administration
Department of Commerce
Room 6013
Washington, DC 20230

National Science Foundation
1800 G St., NW
Washington, DC 20550

National Wildlife Federation
1400 16th St., NW
Washington, DC 20036

Natural Resources Defense
Council
1350 New York Ave., NW
Washington, DC 20005

Renew America
Suite 710
1400 16th St., NW
Washington, DC 20036

The Sierra Club
330 Pennsylvania Ave., SE
Washington, DC 20003

The Wilderness Society
1400 I St., NW
Washington, DC 20005

U.S. Department of Energy
Public Inquiries
Division of Public Affairs
Washington, DC 20585

World Resources Institute
1709 New York Ave., NW
Washington, DC 20006

World Wildlife Fund/
    the Conservation Foundation
1250 24th St., NW
Washington, DC 20037

Zero Population Growth
1400 16th St., NW
Suite 320
Washington, DC 20036

In Canada

Canadian Nature Federation
75 Albert St.
Ottawa, Ont. K2A 1C4

Canadian Wildlife Federation
1673 Carling Ave.
Ottawa, Ont. K2A 3Z1

The Conservation Council
    of Canada
74 Victoria St., Suite 202
Toronto, Ont. M5C 2A5

Energy Resources Conservation
640 5th Ave., SW
Calgary, Alta. T2P 3G4

Environment Canada
Environmental Protection Service
Ottawa, Ont. K1A 1C8

Pollution Probe Foundation
12 Madison Ave.
Toronto, Ont. M5R 2S1

Sierra Club of Eastern Canada
229 College St., Suite 303
Toronto, Ont. M5T 1R4

Sierra Club of Western Canada
620 View St., Suite 314
Victoria, B.C. V8W 1J6

World Wildlife Fund Canada
60 St. Clair Ave. E, Suite 201
Toronto, Ont. M4T 1N5

SELECTED READINGS

American Council for an Energy Efficient Economy (ACEEE). *The Most Energy Efficient Appliances*. Washington, DC: ACEEE, 1988.

Barth, Michael C., and James G. Titus. *The Greenhouse Effect and Sea Level Rise*. New York: Van Nostrand Reinhold, 1984.

Bleviss, Deborah Lynn. *The New Oil Crisis and Fuel Economy Technologies*. Westport, CT: Quorum, 1988.

Bleviss, Deborah Lynn, and Alisa Ayn Gravitz. *Energy Conservation and Existing Rental Housing*. Washington, DC: Energy Conservation Coalition, 1984.

Books on Demand. *The World of Ideas: Essays on the Past and Future*. Ann Arbor, MI: University Microfilms International, 1989.

Broecker, Wallace. "The Biggest Chill." *Natural History*, vol. 96, no. 10, October 1987, pp. 74–82.

Brown, Lester. *The Changing World Food Prospect: The Nineties and Beyond*. Worldwatch Paper 85. Washington, DC: Worldwatch Institute, 1988.

Brown, Lester R., et al. *State of the World 1988*. New York: W.W. Norton, 1988.

Brown, Lester R., et al. *State of the World 1989*. New York: W.W. Norton, 1989.

Brunner, John. *The Sheep Look Up*. (Science fiction.) New York: Ballantine Books, 1972.

Bryson, Reid, and Thomas Murray. *Climates of Hunger: Mankind and the World's Changing Weather*. Madison, WI: University of Wisconsin Press, 1977.

Bunyard, Peter, and Edward Goldsmith, eds. *Gaia: The Thesis, the Mechanisms and the Implications*. Proceedings of the First Annual Camelford Conference on the Implications of the Gaia Hypothesis. Cornwall, England: Wadebridge Ecological Center, Camelford, 1989.

Callenbach, Ernest. *Ecotopia*. New York: Bantam Books, 1970.

Callenbach, Ernest. *Ecotopia Emerging*. Berkeley, CA: Banyan Tree, 1986.

Callenbach, Ernest. *The Ecotopian Encyclopedia*. Berkeley, CA: And/Or Press, 1980.

Center for the Study of Democratic Institutions. "The Shadow Our Future Throws." *New Perspectives Quarterly*, Spring 1989, pp. 2–27.

Clark, Wilson. *Energy for Survival: A New Era*. Garden City, NJ: Anchor Press/Doubleday, 1974.

Committee on Global Change. "Toward an Understanding of Global Change: Initial Priorities for U.S. Contributions to the International Geosphere-Biosphere Program." Washington, DC: National Academy Press, 1988.

Courrier, Kathleen, ed. *Life After '80: Environmental Choices We Can Live With*. Andover, MA: Brick House Publishing Company, 1980.

Crampton, Norman. *Complete Trash: The Best Way to Get Rid of Practically Everything Around the House*. New York: M. Evans, 1989.

Ehrenfeld, David. *The Arrogance of Humanism*. New York: Oxford University Press, 1978.

Ehrlich, Paul, and Robert Ornstein. *New World, New Mind*. New York: Doubleday, 1989.

Flavin, Christopher. "The Heat Is On," *Worldwatch Magazine*, November/ December 1988, pp. 10–20.

Freeman, David S. *Energy: The New Era*. New York: Walker & Co., 1974.

Gever, John, et al. *Beyond Oil: The Threat to Food and Fuel in the Coming Decades*. (A project of Carrying Capacity, Inc.) Cambridge, MA: Ballinger, 1986.

Goldemberg, José, et al. *Energy for a Sustainable World*. Washington, DC: World Resources Institute, 1987.

Goleman, Daniel. *Vital Lies, Simple Truths*. New York: Simon & Schuster, 1988.

Government Institutes, Inc. "Preparing for Climate Change: Proceedings of the First North American Conference on Preparing for Climate Changes, October 27–29, 1987." Washington, DC: Government Institutes, 1988.

Gribbon, John, ed. *Climatic Change*. New York: Cambridge University Press, 1978.

Gribbon, John. *The Hole in the Sky*. New York: Bantam Books, 1988.

Hansen, James, and Sergej Lebedeff. "Global Trends of Measured Surface Air Temperature." *Journal of Geophysical Research*, vol. 92, no. D11, 20 November 1987, pp. 13345–13372.

Harnik, Peter. "How to Live Ecologically in an Apartment." *Environmental Action*, September 1982.

Hayes, Denis. *Rays of Hope: The Transition to a Post-Petroleum World*. New York: W.W. Norton, 1977.

Henderson, Hazel. *The Politics of the Solar Age: Alternatives to Economics*. New York: Anchor Press/Doubleday, 1981.

Hindle, Brooke, and Steven Lubar. *Engines of Change: The American Industrial Revolution 1790–1860*. Washington, DC: Smithsonian Institution Press, 1986.

Hinrichsen, Don. *Our Common Future: A Reader's Guide*. London: Earthscan Books, 1987.

International Geosphere-Biosphere Programme (IGBP). *Global Change: The International Geosphere-Biosphere Programme, a Study of Global Change*. A report prepared by the special committee for the IGBP for discussion at the first meeting of the scientific advisory council for the IGBP, Stockholm, Sweden, 24–28 October 1988. Stockholm, Sweden: IGBP, 1988.

Jager, Ronald. *Eighty Acres*. Boston: Beacon Press, 1990.

Johnston, Robin. "Can Society Act Fast Enough to Save the Environment?" *Christian Science Monitor*, 14 September 1988.

Kerr, Richard. "The Global Warming Is Real." *Science*, vol. 243, no. 4891, 3 February 1989, p. 603.

Kellogg, William, and Robert Schware. *Climate Change and Society: Consequences of Increasing Atmosphere Carbon Dioxide*. Boulder, CO: Westview Press, 1981.

Kornhauser, Andre. "Implications of Ultraviolet Light in Skin Cancer and Eye Disorders." In *Coping with Climate Change*. Washington, DC: The Climate Institute, 1989.

Kumar, Krishan. *Utopia and Anti-Utopia in Modern Times*. New York: Basil Blackwell, 1987.

Laird, Jean. *Homemaker's Book of Energy Savers*. New York: Bobbs-Merrill Co., 1982.

Lappé, Frances Moore. *Rediscovering America's Values*. New York: Ballantine Books, 1989.

Leatherman, Stephen. *Barrier Island Migration: An Annotated Bibliography*. Monticello, IL: Vance Bibliographies, 1985.

Lovelock, James. *Ages of Gaia: Biography of Our Living Earth*. New York: W.W. Norton, 1988.

Lovins, Amory B. "The State of the Art: Lighting." Technical report from Competitek, an information service of Rocky Mountain Institute. Snowmass, CO: Rocky Mountain Institute, 1988.

MacDonald, Gordon. *Climate Change and Acid Rain*. McLean, VA: Mitre Corporation, 1986.

Machado, Sheila, and Rick Piltz. *Reducing the Rate of Global Warming: The States' Role*. Washington, DC: Renew America, 1988.

MacKenzie, James J. *Breathing Easier: Taking Action on Climate Change, Air Pollution, and Energy Insecurity*. Washington, DC: World Resources Institute, 1988.

MacKenzie, James J., and Mohamed T. El-Ashry, eds. *Air Pollution's Toll on Forests and Crops*. New Haven: Yale University Press, 1990.

MacKenzie, James J., and Mohamed T. El-Ashry. *Ill Winds: Air-Borne Pollution's Toll on Trees and Crops*. Washington, DC: World Resources Institute, 1988.

Messel, Harry, ed. *Energy for Survival*. Elmsford, NY: Pergamon, 1979.

Mintzer, Irving. "Weathering the Storms in a Warming World." *Public Power*, November/December 1988, vol. 46, no. 6, pp. 15–21.

Moomaw, William R. "In Search of the Greenhouse Fingerprint." *Orion*, Winter 1989, pp. 5–11.

Myers, Norman, and staff. *Gaia: An Atlas of Planet Management*. New York: Doubleday, 1984.

National Science Foundation. "Global Change." *Mosaic*, vol. 19, Fall/Winter 1988, pp. 1–112.

Nitze, William A. "The Intergovernmental Panel on Climate Change." *Environment*, January/February 1989, vol. 3, no. 1, pp. 44–45.

Norman, Colin. *The God That Limps*. New York: Norton, 1981.

Ogden, Joan M., and Robert H. Williams. *Solar Hydrogen: Moving Beyond Fossil Fuels*. Washington, DC: World Resources Institute, 1989.

Ornstein, Robert, and Paul Ehrlich. *New World New Mind: Moving Toward Conscious Evolution*. New York: Doubleday, 1989.

Pawley, Martin. *Building for Tomorrow: Putting Waste to Work*. San Francisco: Sierra Books, 1982.

Perkins-Schmidt, Drusilla. "Beyond the Almighty Car." *Environmental Action*, vol. 21, no. 1, July/August 1989, pp. 17–26.

Presbyterian Eco-Justice Task Force. *Keeping Justice and Healing the Creation*. Louisville, KY: Presbyterian Church, U.S.A. 1989.

Prindle, William R., and Michael W. Reid. *Making Housing More Affordable through Energy Efficiency*. Washington, DC: Alliance to Save Energy, 1988.

Ramage, Janet. *Energy: A Guidebook*. New York: Oxford University Press, 1983.

Ramirez, Anthony. "A Warming World: What It Will Mean." *Fortune*, 4 July 1988.

Renner, Michael. *Rethinking the Role of the Automobile*. Worldwatch Paper 85. Washington, DC: Worldwatch Institute, 1988.

Repetto, Robert. "Population, Resources, Environment: An Uncertain Future," *Population Bulletin*. July 1987, vol. 42, no. 2, pp. 1–43.

Roberts, Walter Orr. "It Is Time to Prepare for Global Climate Changes." *Conservation Foundation Letter*, 1983.

Russell, Dick. "The Endless Summer." *In These Times*, 11 January 1989, vol. 13, no. 8, pp. 10–13.

Russell, Dick. "Reagan's Legacy of Hot Air." *In These Times*, 25 January 1989, vol. 13, no. 10, pp. 10–13.

Russell, Dick. "Earth Needs You." *In These Times*, 18 February 1989, vol. 13, no. 12, pp. 7–11.

Sale, Kirkpatrick. *Human Scale*. New York: Coward, McCann & Geoghegan, 1980.

Schipper, Lee, et al. *Coming in from the Cold: Energy-Wise Housing in Sweden*. Cabin John, MD: Seven Locks Press, 1985.

Schneider, Stephen. *The Genesis Strategy*. New York: Plenum, 1976.

Schneider, Stephen, and Randi Londer. *The Coevolution of Climate and Life*. San Francisco: Sierra Books, 1984.

Schumacher, E.F. *Small Is Beautiful: Economics As If People Mattered*. New York: Harper and Row, 1975.

Shabecoff, Phil. "Global Warming: Experts Ponder Bewildering Feedback Effects." *New York Times*, 17 January 1989.

Shames, Lawrence. *The Hunger for More*. New York: Times Books, 1989.

Speth, James Gustave. "Environmental Pollution: A Long-Term Perspective." Reprinted by World Resources Institute, Washington, DC, from *Earth 88: Changing Geographic Perspectives*. Washington, DC: National Geographic Society, 1988.

Stanfield, Rochelle L. "Greenhouse Diplomacy." *National Journal*, 4 March 1989, pp. 510–513.

Stobaugh, Robert, and Daniel Yergin, eds. *Energy Future*. New York: Random House, 1982.

Streiber, Whitley, and James Kunetka. *Nature's End: The Consequences of the Twentieth Century*. (Science fiction.) New York: Warner Books, 1986.

Swedish Council for Building Research. *Cost-Efficiency in Custom Design: The Swedish Factory-Crafted Home*. Stockholm: Swedish Council for Building Research, 1989.

Swedish Council for Building Research. *Energy Answers '87: Questions and Answers on Energy Conservation and Management in Buildings and the Built Environment*. Stockholm: Swedish Council for Building Research, 1987.

Swedish Council for Building Research. *Energy in the Built Environment: The Way Forward to the 1990s*. Stockholm: Swedish Council for Building Research, 1988.

Tickell, Crispin. *Climate Change and World Affairs*. Lanham, MD: Harvard University and the University Press of America, 1986.

Titus, James. *Greenhouse Effect, Sea Level Rise and Coastal Wetlands.* Washington, DC: United States Environmental Protection Agency, 1988.

Todd, John, and Nancy Jack Todd. *The Village as Solar Ecology: Proceedings of the New Alchemy/Threshold Generic Design Conference.* East Falmouth, MA: The New Alchemy Institute, 1980.

Todd, John, with George Tukel. *Rehabilitating Cities and Towns: Designing for Sustainability.* San Francisco: Planet Drum Foundation, 1981.

Tolston, Holmes. *Environmental Ethics.* Philadelphia: Temple University Press, 1988.

United Nations Environment Programme (UNEP). *The Greenhouse Gases.* UNEP/GEMS [Global Environmental Monitoring System] Environment Library No. 1. Nairobi: UNEP, 1987.

UNEP. *The Ozone Layer.* UNEP/GEMS Environmental Library No. 2. Nairobi: UNEP, 1987.

United States Environmental Protection Agency (U.S.EPA). *Effects of Changes in Stratospheric Ozone and Global Climate.* Proceedings of the international conference on health and environmental effects of ozone modification and climate change sponsored by the U.S.EPA and the United Nations Environment Programme. Washington, DC: U.S. EPA, 1986.

U.S.EPA. "The Greenhouse Effect: How It Can Change Our Lives." *EPA Journal,* January/February 1989, pp. 2–50.

U.S.EPA. "Our Fragile Atmosphere: The Greenhouse Effect and Ozone Depletion." *EPA Journal,* December 1986, pp. 2–31.

U.S.EPA. *Policy Options for Stabilizing Global Climate.* Washington, DC: U.S.EPA, Office of Policy, Planning and Evaluation, 1989.

United States Government Printing Office (U.S.GPO). "Global Climate Changes: Greenhouse Effect." Hearing before the U.S. House of Representatives Subcommittee on Human Rights and International Organizations of the Committee on Foreign Affairs, 10 March 1988. Washington, DC: U.S.GPO, 1988.

U.S.GPO. "Global Environmental Change Research." Hearing before the U.S. Senate Subcommittee on Science, Technology, and Space and the National Ocean Policy Study of the Committee on Commerce, Science and Transportation, 16 July 1987. Washington, DC: U.S.GPO, 1987.

U.S.GPO. "Global Warming." Hearing before the U.S. Senate Subcommittee on Toxic Substances and Environmental Oversight of the Committee on Environment and Public Works, 10 December 1985. Washington, DC: U.S.GPO, 1986.

U.S.GPO. "Greenhouse Effect and Global Climate Change." Hearing before the U.S. Senate Committee on Energy and Natural Resources, 23 June 1988. Washington, DC: U.S.GPO, 1988.

U.S.GPO. "Greenhouse Effect and Global Climate Change." Hearings before the U.S. Senate Committee on Energy and Natural Resources, 9, 10 November 1987. Washington, DC: U.S.GPO, 1988.

U.S.GPO. "Modeling Greenhouse Climate Effects." Statement of James E. Hansen (Director, NASA Goddard Institute for Space Studies) before U.S. Senate Subcommittee on Science, Technology and Space of the Committee on Commerce, Science and Transportation, 8 May 1989. Washington, DC: U.S.GPO, 1989.

U.S.GPO. "Ozone Depletion, the Greenhouse Effect, and Climate Change." Hearings before the U.S. Senate Subcommittee on Environmental Pollution of the Committee on Environment and Public Works, 10, 11 June 1986. Washington, DC: U.S.GPO, 1986.

U.S.GPO. "Ozone Depletion, the Greenhouse Effect, and Climate Change." Joint Hearing before the U.S. Senate Subcommittee on Environmental Protection, Hazardous Wastes and Toxic Substances of the Committee on Environment and Public Works, 28 January 1987. Washington, DC: U.S.GPO, 1987.

U.S.GPO. "The Potential Impact of Global Warming on Agriculture." Hearing before the U.S. Senate Committee on Agriculture, Nutrition, and Forestry, 1 December 1988. Washington, DC: U.S.GPO, 1989.

"Visions," Environmental Action, May/June 1985, vol. 16, no. 7, pp. 2–17.

Walker, Christopher J., et al. The Impact of Global Climate Change on Urban Infrastructure. Washington, DC: The Urban Institute, 1988.

Wicker, Tom. "A Decade of Decision." New York Times, 28 February 1989.

Winner, Langdon. The Whale and the Reactor. Chicago: University of Chicago Press, 1986.

World Commission on Environment and Development. From One Earth to One World: An Overview. New York: Oxford University Press, 1987.

World Commission on Environment and Development. *Our Common Future*. New York: Oxford University Press, 1987.

World Meteorological Organization and the United Nations Environment Programme. "Developing Policies for Responding to Climatic Change." A summary of the workshops held 28 September–2 October 1987 and 9–13 November 1987 under the auspices of the Beijer Institute, Stockholm, Sweden. Geneva, Switzerland: World Meteorological Organization and UNEP, 1988.

World Resources Institute and the International Institute for Environment and Development. *World Resources 1988–89*. New York: Basic Books, 1988.

Yepsen, Roger. *The Durability Factor*. Emmaus, PA: Rodale, 1982.

About the Authors

Francesca Lyman, a journalist for 10 years, has been editor of *Not Man Apart, Environmental Action,* and the national feature service Words by Wire. She has written for the *New York Times, New Jersey Monthly, New Age, Progressive, Mother Jones, Special Reports,* and *Technology Review.* Ms. Lyman lives in Bronxville, New York, with husband Bob Aglow and son Devin.

Irving M. Mintzer is a senior associate in the Climate, Energy, and Pollution Program at the World Resources Institute, Washington, D.C. Dr. Mintzer has served as a member of the U.S. Environmental Protection Agency's Science Advisory Board on the Assessment of the Risks of Stratospheric Modification; as an advisor on STARPOWER: The U.S. and the International Quest for Fusion Energy; and as a consultant to the International Institute for Applied Systems Analysis in Austria.

Kathleen Courrier directs the World Resources Institute's publishing program. Ms. Courrier has been acting director of the Center for Renewable Resources, a publishing consultant at the Organization of American States, and the editor at Worldwatch Institute. Her articles and reviews have appeared in the *Christian Science Monitor, Columbia Journalism Review, Sierra,* the *Washington Post,* and other periodicals, and she is the editor of *Life After '80: Environmental Choices We Can Live With.*

James J. MacKenzie is a senior associate in the World Resources Institute's Program in Climate, Energy, and Pollution. A physicist, Dr. MacKenzie was formerly senior staff scientist at the Union of Concerned Scientists and senior staff member for energy at the President's Council on Environmental Quality. His publications include *Ill Winds: Airborne Pollution's Toll on Trees and Crops* and *Breathing Easier: Taking Action on Climate Change, Air Pollution, and Energy Insecurity.*

INDEX

187

About WRI

World Resources Institute (WRI) is an independent research and policy center in Washington, D.C., that helps governments, other nonprofit organizations, businesses, and citizens look for ways to meet basic human needs and nurture economic growth without degrading our planet and its resources.

The institute's staff of scientists, economists, political scientists, communicators, and others collaborate in five broad programs: climate, energy, and pollution; economics and institutions; resource and environmental information; forests, biodiversity, and sustainable agriculture; and technical assistance in resource assessment, planning, and management.

WRI is funded by private foundations, the United Nations, international government agencies, corporations, and concerned individuals.

World Resources Institute
1709 New York Ave., NW
Washington, DC 20006